Ripples in the Cosmos

The most luminous galaxy in the universe (see page iv for full caption)

RIPPLES IN THE COSMOS

*A view behind the scenes
of the new cosmology*

MICHAEL ROWAN-ROBINSON

W.H. FREEMAN
SPEKTRUM

OXFORD · NEW YORK · HEIDELBERG

W.H. Freeman and Company Limited
20 Beaumont Street, Oxford OX1 2NQ
41 Madison Avenue, New York, NY 10010

British Library Cataloguing-in-Publication Data.
A catalogue record for this book is available from the British Library.

Library of Congress Cataloging in Publication Data

Rowan-Robinson, Michael.
Ripples in the Cosmos: a view behind the scenes of the new cosmology/
by Michael Rowan-Robinson.
p. cm.
Includes bibliographical references and index.
ISBN 0-7167-4503-8
1. Cosmic ripples. 2. Cosmology. 3. Cosmic background radiation.
4. Infrared Astronomical Satellite. I. Title
QB991. C64R68   1993
523. 1– dc20   93-1590
CIP

Set by Keyword Publishing Services Ltd.
Printed by The Bath Press Ltd.

Frontispiece

The most luminous galaxy in the universe, IRAS F10214+4724 (see chapter 13), imaged in
infrared light with the new 8-meter diameter Keck Telescope on Mauna Kea, Hawaii. This is
the first scientific image generated with the telescope and was shown at the American
Astronomical Society meeting in June 1993. The galaxy, with a luminosity of 500 million
million suns, 99% of which is emitted at infrared wavelengths, was discovered by the author
and his colleagues during a ground-based observing programme following up the all-sky survey
made by the IRAS Infrared Astronomical Satellite. This Keck Telescope image shows that
the galaxy (centre) appears to consist of several fragments in the process of merging together.

In affectionate memory of
Michael Penston

Contents

Prologue

What kind of a universe?

WHEN WE LOOK at the earth with its teeming life, its complex land-scape of continent and ocean, it is natural for us to assume this is a typical environment in the universe. Yet as soon as we study the solar system, the sun and its retinue of diverse planets, we find that there are immense spaces between the planets. And looking up at the night sky, peering at the feeble spots of light that are the stars scattered across the profound blackness, the vast emptiness of the universe begins to come home to us.

It is a curious fact that the average densities of the earth, the sun and the planets, are not very different from that of the human body. Our bodies have an average density close to that of water – 1 gram per cubic centimetre. The average densities of the planets range from 0.7 grams per cubic centimetre for Saturn to 5.5 for the earth, and the sun's mean density is 1.4 grams per cubic centimetre. So this seems like a very characteristic and familiar scale of density.

Yet when we calculate the average density of the universe today, adding up all the matter in the stars, the galaxies, the clusters of galaxies, we find an almost incomprehensibly low figure, about a million million million million million times lower than that of water (10^{-30}

1

grams per cubic centimetre). So the contrast between the regions of the universe where matter is gathered into stars, planets or human beings, and the almost perfect vacuum of the space between, could hardly be stronger.

The universe today displays these extraordinary contrasts of density on the small scale. But when we search to ever larger scales the picture gradually changes. We find that the stars are gathered in vast diffuse systems, the galaxies, typically a hundred thousand light years across. The Milky Way, that hazy band of starlight across the sky, traces the plane to which the stars of our own Galaxy are concentrated. In turn the galaxies are aggregated into groups and clusters which may be as large as tens of millions of light years in size. On these scales the contrasts in density between one region of the universe and another are much less dramatic, no more than a factor of ten or so. We have come to realize in the past few decades that on the very largest scales that we can measure, the universe is almost unbelievably smooth and uniform. This smoothness and uniformity is a source of great perplexity for cosmologists, as we shall see, and they have resorted to some strange explanations of how this state of affairs came about.

In 1929, the year of the Wall Street Crash and of the beginning of the Great Depression in Europe and the United States, Edwin Hubble discovered that the universe is expanding. Psychologically, this is a disturbing idea, suggesting an impermanence, even a transience about the universe. Did this mean the universe had a sudden explosive beginning in a 'Big Bang', about 10 to 15 billion years ago, or could this expansion have continued for ever in a 'steady state'? The nature of the universe's infancy became clearer with the discovery in 1965 by Arno Penzias and Robert Wilson of background radiation at microwave wavelengths. This was immediately interpreted as relic radiation from the 'fireball' phase of a Big Bang universe, an explanation which has been reinforced by subsequent studies. We are getting a snapshot of the universe as it was only a few hundred thousand years after the birth of the universe in the Big Bang. The microwave background radiation was found to be extraordinarily smooth, the same in every direction on the sky to an accuracy of better than 0.01 per cent, once the effect of our Galaxy's motion through space was corrected for.

How did the universe evolve from such a state of almost perfect uniformity to the vast contrast in densities we see today? This question of how structure arose in the universe has been at the heart of the cosmological debate for the past decade. In the past few years we have taken two enormous steps forward. Both have involved telescopes in orbit around the earth working at wavelengths invisible to the human eye. The first step has been the mapping of the distribution of galaxies in the universe on hitherto unprecedentedly large scales, using galaxies detected in a survey of the sky made by the Infrared Astronomical Satellite (IRAS). The second step was the detection of minute 'ripples' in the microwave background radiation by the Cosmic Background Explorer (COBE). The purpose of this book is to describe these discoveries and how they are connected, and to discover how they help us solve the problem of how structure, and hence we ourselves, evolved in the universe. This is a story which links the most fundamental ideas about the structure of the atomic nucleus, elementary particle physics, to the origin and evolution of the whole universe.

The IRAS story

A major part of this book is the inside story of the Infrared Astronomical Satellite, known as IRAS – the seven years of preparation by American, Dutch and British astronomers and space scientists before the launch of IRAS by NASA in 1983, the ten hectic months in 1983 when it mapped the far infrared sky for the first time, and the subsequent decade of follow-up with some of the world's greatest ground-based telescopes.

It is a personal view of that story and it is, therefore, a very partial and incomplete view. I saw little of the building of the spacecraft and its revolutionary telescope. I therefore pass rather lightly over the immense problems faced and solved by the hundreds of men and women who built IRAS: someone else will have to write a better version of their story.

I was quite closely involved with the analysis of the IRAS data and the preparation of the catalogues of hundreds of thousands of infrared sources seen by IRAS, so I am on more solid ground with that part of the story. And with my team of researchers at Queen Mary and Westfield College, London, and collaborators at other universities, I was involved in some very exciting programmes following up the IRAS survey with ground-based telescopes, which led to some dramatic discoveries. Some of these discoveries provoked stories in the newspapers and on radio and television. One point of this book is to try to show what the real story behind those media articles and reports was.

To explain the IRAS story and its contribution to our understanding of the growth of structure in the universe, I have to tell you quite a lot about infrared astronomy and cosmology, but I have tried to confine this to what I think you need to know to follow the main story. This means that this is also a very partial and incomplete account of infrared astronomy and cosmology. Some people who work in these fields get mentioned, many do not: perhaps only as few as one-tenth of the people who ought to get mentioned, do.

The microwave background radiation

The IRAS story is intimately connected to that of the microwave background radiation, so I have also given the full story of the discovery of this radiation, and how it was explained as the relic of the early stages of a Hot Big Bang universe. This story was a bizarre catalogue of near-misses and misunderstandings, before Arno Penzias and Robert Wilson in 1965 finally made the discovery which was to transform cosmology.

At first, the most striking feature of the microwave background was its isotropy, the fact that it looks the same whichever direction you look in the sky. Einstein had suggested a simple isotropic model for the universe in 1917, but even in 1965 few would have believed how good an approximation this was to turn out to be. Year by year the accuracy with which this isotropy was measured improved. First 10%, then 1%, then 0.1%. Then in 1977, twelve years after the discovery of the microwave background radiation, the first departure from perfect isotropy was found. Maps of the radiation round the sky by several groups showed that it was slightly brighter than average in one direction, by about 0.1%, and slightly dimmer than average, by the same amount, in the opposite direction. The explanation was that our Galaxy is moving through space at an incredible speed of 600 kilometres per second.

For many years the origin of this motion remained a puzzle. Then in 1990, three-dimensional maps of the distribution of IRAS galaxies in a large volume of the universe made by my colleagues and I allowed us to explain this rapid motion through space, in terms of the gravitational pull of clusters of galaxies within 300 million light years of earth. Our IRAS galaxy maps also gave us a measurement of the average density of matter in the universe which was at least ten times higher than that of the matter we can see in galaxies. This was concrete evidence that the universe is filled with some kind of dark matter.

The second dramatic discovery from our IRAS galaxy surveys, published in 1991, was that the universe was 'lumpier' on larger scales than expected according to the then current ideas about galaxy formation. This large-scale structure we were finding has a crucial bearing on how galaxies and clusters of galaxies formed, and on the nature of the dark matter in the universe.

The ripples

We come now to the story of the discovery of the 'ripples' in the microwave background radiation, found by The National Aeronautics and Space Administration's (NASA's) COBE satellite in 1992. The image of ripples is quite a good one. When we stare at the surface of a pond or lake disturbed by wind, we see the main wave motion driven by the wind. But we also see running away in every direction smaller irregularities, the ripples, which are driven by the surface tension of the water. We see a hierarchy of disturbance, from the largest waves down to the smallest ripple. In the universe too we see a hierarchy of disturbance from large scales to small scales. The difference, though, is that on the surface of a lake, the disturbances of largest amplitude are the metre-scale waves, with the most miniscule ripples being almost imperceptible. In the universe, it is on the scale of stars that we see the greatest contrast in density. As we go through the hierarchy increasing the scale from galaxies, through clusters of galaxies, to the very largest structures detectable in the universe, the amplitude of the density contrast decreases. On the gigantic scale probed by COBE the density contrast is a minute fraction of a per cent, a mere ripple on the smooth background radiation.

The announcement of COBE's discovery of the ripples burst on most of us like a bolt from the blue and astronomers did their best to explain the story to the excited journalists. Even though I was not involved in the discovery, this was a very exciting time. Although there were exaggerations in some of the reports, the scale of the coverage was appropriate to the magnitude of the discovery. My view is that science needs more of this kind of story and this kind of coverage, not less. Considering the importance of science in our lives, the space devoted to science in the newspapers and media ranges from minimal to pathetic.

There was also a very exciting race to come up with an explanation of what the 'ripples' meant for cosmology. Gradually it dawned on me that the COBE ripples and the IRAS lumps were two ends of the same stick. Tying them together gave the solution to the problem of how galaxies and other structures in the universe formed, and what the nature of the dark matter in the universe had to be. Several different groups of astronomers came to similar conclusions.

As well as telling a particular scientific story this book also gives a picture of what science is, how it works, and what it is like to work in the space programme and with the great modern telescopes. The first and last chapters deal specifically with the philosophical issues of what science is about. Is the universe mathematical in essence? Can the meaning of life be derived from science? By showing scientists at work on a particular programme, I think I show more clearly what science is like than any philosophical discussion can. I also try to respond to some of the current attacks on science.

I have been thinking about writing this book for several years, but each year seems to bring some new exciting twist to the IRAS story. In the latest developments we are pinning down the nature of the dark matter in the universe, testing ideas about the earliest fraction of a second after the Big Bang, and catching galaxies in the moment of their formation. I have been fortunate to be close to some of these dramatic discoveries. The story is not over yet.

The lure of number

Is the universe mathematical in essence? Leonardo wrote in his note-books that no one who was not a mathematician should read his work. Since the time of the Greeks all advances in our understanding of the physical universe have needed mathematics to formulate them. But does mathematics lie at the heart of the structure of the universe, or is it simply the way we human beings grapple with the part of it we can perceive? Many scientists just get on with the business of doing science without worrying too much about philosophical issues of this kind. For others, like myself, the philosophical basis of science, and how science relates to the rest of human culture, is a fundamental issue.

I try to get at this question of whether mathematics is at the heart of the structure of the universe by considering some simple examples from the past. In each case where some branch of mathematics was thought to be fundamental to the nature of the universe, this turned out to be an illusion. Mathematics gives us useful tools, but it is science, through observation, theory and experiment, which gives us a picture of the world. A theory like Einstein's General Theory of Relativity gives us a mathematical picture of reality, which we use with confidence because of its many experimental successes.

7

The fascination of number

Many children find themselves fascinated by the idea of numbers when they first encounter them. Some become intimidated by mathematics later in school and regard it for ever as a closed book. For many people, their terror of mathematics is the main obstacle to their getting to grips with science. I was lucky because my mother was a mathematician and talked to me about numbers when I was very small.

To anyone who is fascinated by them, numbers do not all look the same. Because we have five fingers on each hand there is something very special about the numbers 1 to 5. This is encapsulated in the Roman number system, which goes I, II, III, IIII, V and then continues VI, VII and so on. This is like many primitive runic number systems which change their symbols at every multiple of 5. Of course today we are more used to a number system in which the number 10 plays the key role, the Indian and Arabic system of writing numbers, which is obviously based on the number of fingers of both hands. We say that our number system has base 10. But 10 is a completely arbitrary choice. Computers prefer to work with a system of numbers with base 2, the binary system.

So 10 is a special number for human numerologists. But then so is 12, because after 12 in many languages the numbers change their names to thir-teen, four-teen, etc. Why is 12 special? Clearly because it is the smallest number with the factors 2, 3 and 4, which makes it useful in all practical matters involving dividing into 2, 3 or 4 parts. Hence the appearance of the number 12 in our units of time (the day and night were each originally divided into 12 hours, the length of which would depend on the season), and until recently in the British measures of distance (12 inches in a foot) and of money (12 old pence in a shilling). The smallest number with the factors 2, 3, 4, and 5, the five fundamental numbers of one hand, is 60 and that is another special number. The Babylonians chose to base their number system on 60 and we still have the relics of that today in our measurement of time (60 seconds in a minute, 60 minutes in an hour) and of angle (60 seconds of arc in a minute of arc, 60 minutes of arc in a degree). So these special numbers 1–5, 10, 12, 60 are all linked to our bodies and are embedded in our culture.

Scattered among the numbers are those odd-looking numbers, the primes, which do not have any factors apart from 1 and themselves. The numbers 7, 11, 13, 17, and so on just feel different from their neighbours, you cannot do anything with them. The Greek mathematician Euclid showed that there is no largest prime number, they go on for ever. There used to be a certain fascination in trying to find ever larger prime numbers, and in speculations that certain formulae, like that of the seventeenth century French mathematician Pierre Fermat, $2^{2^n} + 1$, might generate only prime numbers, but the fun of these speculations has been eliminated by computers, which just grind away remorselessly. To make matters worse, large prime numbers now have a military application, in cryptography. Because it is very much harder to find the prime factors of a large non-prime (say of 200 digits) than it is to prove that such a number is prime, it is possible to construct very large numbers which are the products of two large primes. This product can be published and it will be impossible for anyone to find the factors. The product can then be used to code messages, which can only be decoded by someone who knows what the factors are.

The largest primes known have been found by studying the Mersenne numbers, which are of the form $2^p - 1$, where p is a prime number. For $p = 2, 3, 5, 7$ the Mersenne formula gives 3, 7, 31, 127, which are all prime, but there are in fact only 27 Mersenne primes up to $p = 60,000$. The largest of these is $2^{44,497} - 1$, which was found by Harry Nelson and David Slowinski using a Cray computer at Lawrence Livermore Laboratory. The largest prime known, $2^{756,839} - 1$, discovered in 1992, is also a Mersenne number.

The Pythagorean cosmology of number

One of the earliest mathematical cosmologies was that of Pythagoras, based on the natural numbers. Pythagoras, who lived in the sixth century BC, had discovered that musical notes that sounded harmonious together bore a simple arithmetical relation to each other. Thus two notes an octave apart, which clearly resonated together in a special way, would be formed by stringed instruments in which the length of the string in one was exactly twice that of the other. If the lengths are in the ratio of 3 to 2, the chord known as the fifth sounds, and so on for other small whole number ratios. From these wonderful discoveries Pythagoras was inspired to suggest that the whole universe might show such an arithmetical harmony and for many centuries scientists strove to discover the 'music of the spheres'. Pythagoras founded a powerful school of learning, the leadership of which on his death passed to his wife Theano and to his daughters. A principle of his academy was that knowledge was passed from the 'magus' to his or her followers in secrecy. One reason for this secrecy will become clear shortly. This concept of the magus, possessed of secret knowledge from the past, knowing the numerical key to the universe, was a powerful image in renaissance culture.

Fig. 2.1 In The Harmonies of the World *(1619), Johannes Kepler sought to explain the relationship between the size of the orbits of the planets in terms of musical harmonies.*

One of the fascinating historical discoveries of recent years has been the extent to which Isaac Newton (1642–1727), often called the father of the age of reason, saw himself as the last of the magi. It was no accident that the number of colours into which Newton chose to divide the colours of the rainbow, of the visible spectrum, was the mystical seven. For Newton the division of white light into seven constituent colours was a pure Pythagorean insight. From his hermetic and astrological studies he drew inspiration for his idea of a gravitational force acting across space, which Leibniz was to dismiss as 'a senseless occult quality'.

2000 years of Euclidean geometry

From the simple whole numbers we can branch off in many different directions. One of the most natural and useful is geometry. The first great treatise on geometry was that compiled by Euclid in about 300 BC, which summarized several centuries of Greek achievements in geometry and was to have a dominant effect on mathematics, physics and astronomy for 2000 years. It is only one generation ago that Latin and Euclid were still taught in British schools, and that pupils therefore shared the same education, to start with at least, as all the European peoples of the past 500 years. Perhaps it will be harder for students in the future to understand the minds of the thinkers of the past who were immersed in those classical studies.

The theorems on the circle summarized by Euclid provided Plato (428–347 BC) with the basis for a new mathematical cosmology, based on the circle and the sphere. To a good approximation, the shape of the earth is a sphere, the moon and the sun look circular and their orbits round the earth are circular. To Plato the divine nature of the heavens implied that the universe had to be a sphere and the planets had to move in the perfect form of a circle. Thus Plato's cosmology, extended into a detailed scheme by his pupil Aristotle (384–322 BC), accounted for some of the most basic phenomena known about the universe. In a remarkable development of these ideas, Ptolemy in the second century BC was able to account for the motions of the sun, moon and five known planets as an elaborate combination of circular motions, in a model which was to be used for over 1600 years. It is unlikely that any world-view will ever last as long as that again.

Nicholas Copernicus (1473–1543), whom we think of as the founder of modern astronomy, was still rooted in this Platonic vision. Ironically it appears that his discovery that the planets orbit the sun rather than the earth was made in the course of trying to revise Ptolemy's model so that it more perfectly embodied Plato's vision. Ptolemy had been forced to use 'equants', in which the circular motion is about a point which is not the centre of the circle. By forcing all planetary motions to be described by epicycles, regular motions in a circle about a centre which itself moves on a circle, Copernicus found that, viewed from the earth, the same epicycle was present in the motions of all the planets and that this was the opposite motion to that of the sun. He was thus able to simplify the model enormously by placing the sun at the centre of the solar system. However he still needed 34 circular motions in his model to account for the motions of the planets on the sky, so his system was only slightly less complex and cumbersome than Ptolemy's.

It took Johannes Kepler (1571–1630), immersed in the geometry of the regular solids and conic sections, to realize that the whole apparatus of epicycles could be replaced by simple motion in an ellipse. Yet Kepler too was motivated by the Pythagorean goal of projecting a har-

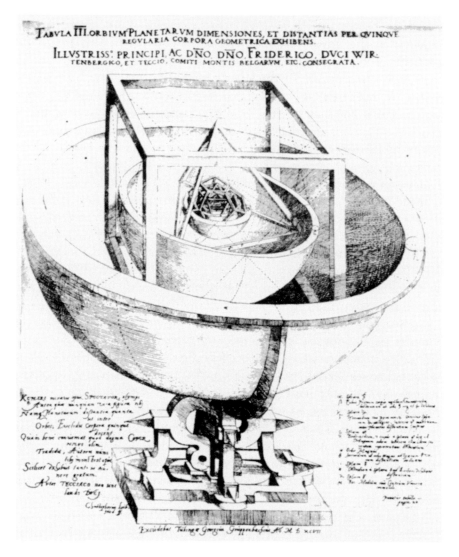

Fig 2.2 Kepler's explanation of the structure of the solar system using the five regular polyhedra nested between the spheres of the planets. From The Harmonies of the World *(1619).*

monious mathematics onto the universe and he sought to embed the planetary orbits in a nested sequence of regular solids. Thus in each of the cases of Pythagoras, Plato and Kepler, we see a brilliant success in using a piece of mathematics (harmonics, circular motion, and the geometry of the ellipse) to describe some fundamental aspect of the universe coupled with a completely fallacious attempt to force this mathematics on everything.

The philosopher Immanuel Kant (1724–1804) was completely fooled by the apparently fundamental nature of Euclidean geometry and postulated that the properties of Euclidean space were given *a priori*, that is they are embedded in the structure of the universe and can be assumed as something given. But with the hindsight of Einstein's General Theory of Relativity, we can see that this view is profoundly mistaken. The first cracks in the stranglehold of Euclidean geometry on the human imagination appeared in 1826 when the Russian mathematician Nikolas Lobatchewsky invented non-Euclidean geometry, in which parallel lines can meet and the angles of a right-handed triangle do not add up to 180°.

Pythagoras discovers the fatal flaw in his cosmology

One of the most important results in Euclid's compilation is Pythagoras' theorem: in a right-angled triangle the square on the hypotenuse is equal to the sum of the squares on the other two sides. So if the other two sides both have length 1 (in any units you choose), the length of the hypotenuse will be the square root of 2. Pythagoras was horrified to realize that $\sqrt{2}$ could not be written as the ratio of two whole numbers n/m, say, because he had based his whole cosmology on the whole numbers. Pythagoras and his followers tried to suppress this result but truth, as they say, will out. $\sqrt{2}$ is what we now call an *irrational* number. This is not because it defies reason (it obviously does not) but because numbers like n/m, where n and m are whole numbers, are called *rational* numbers (they are ratios). Irrational numbers are defined to be those that cannot be written as a ratio of whole numbers. And this means their decimal expression goes on for ever without recurring. Another very important irrational number is π, the ratio of the circumference of a circle to its radius. Approximations to this number were in widespread use as early as 2000 BC. Pythagoras' reaction to his discovery is typical of mathematicians, who tend to think that mathematics rules the universe. Having based his cosmology on a particular area of mathematics, whole numbers, he was not going to let a little bit of inconsistency with the real world spoil it.

Another area of mathematics which opens off from the theory of numbers is algebra. If x, y, z are the lengths of the three sides of a right-angled triangle, Pythagoras's Theorem can be written in algebraic form

$$x^2 + y^2 = z^2$$

This is the basis of a whole version of geometry invented by the French mathematician and philosopher René Descartes in the seventeenth century. All the results of Euclid can then be proved purely algebraically. Moreover, this algebraic geometry can be extended to more than three dimensions or to curved space. This step was eventually taken by the German mathematician Bernhard Riemann in 1854.

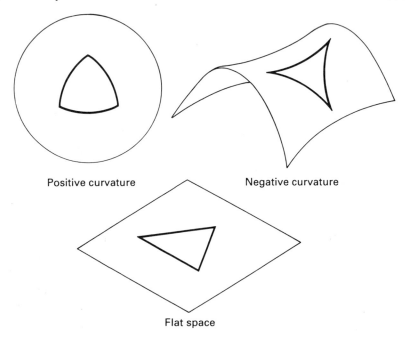

Positive curvature   Negative curvature

Flat space

Fig 2.3 Examples of non-Euclidean geometry: spaces of positive curvature (top left) and negative curvature (top right), compared with a flat Euclidean space (bottom).

Riemann's algebraic geometry of curved, four-dimensional space was to be just the mathematical tool Einstein was looking for in his attempts to develop a relativistic theory of gravitation and formed the basis of his 1916 General Theory of Relativity.

An aside on Fermat's last theorem

An area of number theory which I have always founded fascinating is a 'theorem' which still has not been proved after three centuries of effort.

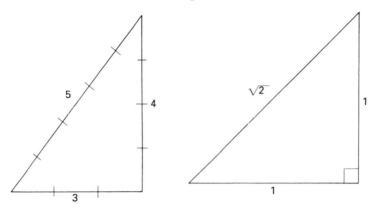

Fig. 2.4 The 3-4-5 and 1-1-√2 triangles.

A solution of the Pythagorean equation above, which has been known since remote antiquity, is $x = 3$, $y = 4$, $z = 5$. Obviously, any multiple of this set of numbers would be a solution and there are infinitely many other sets of whole number solutions. The seventeenth-century French mathematician Pierre de Fermat announced a very interesting speculation just before he died, which has become known as *Fermat's last theorem:*

The equation $x^n + y^n = z^n$ has no whole number solutions for $n > 2$.

The theorem was found written in the margin of one of Fermat's textbooks with the comment 'I have discovered a truly marvellous demonstration [of this theorem] which this margin is too narrow to contain.' Fermat did leave a proof of the case $n = 4$, and the cases $n = 3$, 5 and 7 were proved by the famous mathematicians Leonhard Euler, Adrien-Marie Legendre and Augustin-Louis Cauchy. In 1847 Ernst Kummer found a condition on the number n which, if satisfied, ensured that Fermat's theorem was true for that n. Kummer's condition implied that Fermat's theorem was true for all n up to 100 except $n = 37$, 59, and 67. Today Fermat's theorem has been verified for all n up to 125,000. But is it definitely true for all n? I became fascinated by this theorem, and other aspects of what is called the 'theory of numbers', when I was working at the National Physical Laboratory, Teddington, before starting university. The reason I chose NPL for this pre-university job was to try to decide whether to study Mathematics or Physics at university. At NPL I had to choose between working in the Mathematics Division, where Alan Turing had built his pioneering ACE computer after the Second World War, or the Light Division. To see what physics was like, I opted for light rather than mathematics, a choice which perhaps set me on the road to a career in astronomy. I worked with the UK's International Standards of Luminosity, again highly appropriate for my later interest in the cosmological distance scale, which involves trying to find standards of luminosity on an extragalactic scale. The International Standards of Luminosity were once candles made in a precisely defined way, hence the phrase 'standard candle' which cosmologists use of a class of astronomical object which they think always has the same luminosity and so can be used to estimate distances. Today these standards are essentially huge light bulbs and each requires many hundreds of hours of calibration. I have to confess that I once dropped one and it smashed to pieces. I was hastily moved to another section where I worked on the calibration of detectors with a huge spectrometer. Interestingly, I was given a new silicon detector to test, a prototype which a manufacturer had asked NPL to evaluate. This detector was unusual in that it was quite sensitive to infrared radiation but it had the annoying habit of changing its response every time you reobserved it and eventually I had to give up trying to calibrate it. Years later during the IRAS mission I realized that I must have stumbled on the phenomenon of 'hysteresis' in this type of infrared detector, where the electrical response of the detector changes when the detector is illuminated by radiation, a problem which plagued the IRAS mission.

In my free time I studied Fermat's last theorem and there was a glorious period of 24 hours when I thought I had proved it. Unfortunately I then found a mistake. Along the way I had become interested in the theory of partitions. The partition of a number n, called $p(n)$, is the number of ways n can be written as a sum of smaller numbers. For example $6 = 5+1 = 4+2 = 4+1+1 = 3+3 = 3+2+1 = 3+1+1+1 = 2+2+2 = 2+2+1+1 = 2+1+1+1+1 = 1+1+1+1+1+1$, so there are 11 ways of writing 6 as a sum of smaller numbers and $p(6) = 11$. Here I had more success and I discovered a way of building up the value of $p(n)$ by means of a triangle (mathematics students will know of a similar triangle discovered by Pascal). I later published this in a student magazine. This was my only and very minor contribution to pure mathematics. Fermat's Last Theorem has still not been proved, though in 1983 Gerd Faltings showed that, for any particular value of n, the number of integer solutions is finite. When compared with the infinite number of possible solutions this seems tantalizingly close to a proof. In retrospect, after the centuries of effort to prove Fermat's last theorem, it seems most unlikely that Fermat really did find a proof of it.

1	1	1	1	1	1	1	1	1	1	1	1	1	1
	1	1	1	1	1	1	1	1	1	1	1	1	1
		1	2	2	2	2	2	2	2	2	2	2	2
			1	2	3	3	3	3	3	3	3	3	3
				1	3	4	5	5	5	5	5	5	5
					1	3	5	6	7	7	7	7	7
						1	4	7	9	10	11	11	11
						0(+1)	1(+3)	4(+4)	8(+3)	11(+2)	13(+1)	14(+1)	15
								1	5	10	15	18	20
									1	5	12	18	23
										1	6	14	23
											1	6	16
												1	7
													1
1	2	3	5	7	11	15	22	30	42	56	77	101	135

Fig. 2.5 A triangle for calculating partitions: the differences along the horizontal rows are given by the terms in the vertical columns, reading off from the diagonal. The partition of n, P(n), is then the sum of the terms of the nth column.

Does mathematics rule the universe?

Fascinating though number theory is, it does not have much application to the study of the universe. Working in a mathematics department I hear a lot of propaganda about the importance of pure mathematics for its own sake. My own view is that although mathematics has intrinsic cultural value and a definite fascination to those with a certain turn of mind (myself included), mathematics only takes on its great importance in human culture as a tool for science, especially for physics. Thus I take the view that mathematics is essentially a tool of theoretical physics, rather than an end in itself. I also think that all claims that the universe is in essence mathematical are an illusion.

As an illustration of this view, let us look again at Pythagoras and the discovery of irrational numbers. In fact Pythagoras, in trying to suppress this result, was overreacting to his discovery. Irrational numbers do not play any part in the real world. If we measure the side of a real right-angled triangle with a ruler, we might come up with the answer 1.000 metres for each of the two sides next to the right angle and 1.414 for the hypotenuse. As we increase the accuracy of the measurements we do not get any closer to proving that the hypotenuse is $\sqrt{2}$. Instead as the accuracy gets greater we find something rather interesting. In the real world Pythagoras's theorem *does not hold* because space-time is in fact slightly curved by the earth's gravity. We would then find that Einstein's General Theory of Relativity would be needed to get the 'right' answer. Eventually, perhaps, we would reach a level of accuracy where even General Relativity would no longer be correct. So far, in modern physics, we have not reached that level of accuracy. But General Relativity, like Euclidean Geometry before it, is one particular mathematical picture of reality. While it works, it is useful and beautiful. But there are always many possible mathematical pictures, many possible geometrical systems. Only one of these will work in the actual world and we can only decide which one by experiment and observation.

The great Austrian philosopher Ludwig Wittgenstein, who lived much of his life in England, stated this point of view forcefully in his *Tractatus*, written in the trenches and barracks of the 1914–1918 war:

The propositions of logic are tautologies.
Therefore the propositions of logic say nothing.

Mathematics is a logical method.
A proposition of mathematics does not express a thought.
Indeed in real life a mathematical proposition is never what we

want. Rather, we make use of mathematical propositions only *in inferences from propositions that do not belong to mathematics to others that likewise do not belong to mathematics.*

It is an hypothesis that the sun will rise tomorrow: and this means that we do not know *whether it will rise.*

The whole modern conception of the world is founded on the illusion that the so-called laws of nature are the explanations of natural phenomena.

These oracular summaries of Wittgenstein's views on the nature of mathematics made a tremendous impression on me when I first read them as a student. Rereading them recently, and Ray Monk's wonderful biography *Ludwig Wittgenstein, the Duty of Genius*, I still find Wittgenstein one of the most inspiring thinkers of the century.

To return to our particular example of Pythagoras's right-angled triangle above, there is no measurement or experiment that can yield as its result an irrational number. Such numbers appear in idealized mathematical pictures like Euclidean geometry and are extremely useful. But it does not follow that the universe has been designed mathematically. Mathematics gives us useful tools, but it is not a science. It is science, through observation, theory, experiment, which gives us a picture of the world, not mathematics.

Bertrand Russell describes in his *Autobiography* how he turned to mathematics hoping to find a certainty there which was absent in life and in the world. He and Alfred Whitehead set out in the early part of this century to construct a logical basis for arithmetic. In the same period David Hilbert was setting out his programme for a self-consistent foundation for mathematics. These heroic enterprises were conclusively shown to be a chimera by the logical work of Kurt Gödel and Alan Turing during the 1930s. An excellent account of this latter work is given in Roger Penrose's *The Emperor's New Mind*. Very briefly, Gödel showed in 1931 that the self-consistency of arithmetic could not be demonstrated within the framework of arithmetic itself. The system is incomplete and contains statements whose truth is undecidable except by appealing to additional, higher-level axioms. In 1937 Turing proved an equivalent result, that there are mathematical expressions which are not computable. These demonstrations that the foundations of mathematics are built on clay do not deter writers on the philosophy of mathematics like Roger Penrose and John Barrow (in *Pi in the Sky*) opting for a Platonic view of mathematics, that

> *mathematical truth is absolute, external and eternal, and not based on man-made criteria; and that mathematical objects have a timeless existence of their own, not dependent on human society nor on particular physical objects.*
>
> (Penrose: *The Emperor's New Mind*)

I remain unconvinced. Mathematics is a brilliant creation of mathematicians and like all human creations, it contains flaws and limitations. To match mathematics to the universe requires a further kind of brilliance, insight into physics and into what is known about the universe. This tends to be a patchwork process which most modern writers on physics and cosmology tend to play down.

What makes a good new scientific theory? Obviously it has to be consistent with the well-established parts of earlier theories, but it must also explain phenomena not previously understood, and should make predictions of new phenomena. This criterion of experimental testability, articulated forcefully by Karl Popper in his *The Logic of Scientific Discovery*, is a vital one for a good scientific theory.

Einstein's General Theory of Relativity of 1916 provides an instructive example. In 1905 Einstein had introduced his Special Theory of Relativity to make the dynamics of moving bodies compatible with the fact, found in the Michelson-Morley experiment of 1887, that the speed of light does not depend on the relative velocity of the source and the observer. This required modifications to dynamics when the speeds of moving bodies approached the speed of light. In his General Theory of Relativity, Einstein went a step further and incorporated gravitation too. Matter has the effect of curving up space-time around it and the path of light becomes a curve.

The theory gives the same results as Newtonian gravity when the gravity is not too strong, so none of the achievements of Newtonian gravitation were thrown away. It accounted for an anomaly in the orbit of Mercury, which had been unexplained for over a century. And it predicted a whole series of subtle phenomena which have been verified

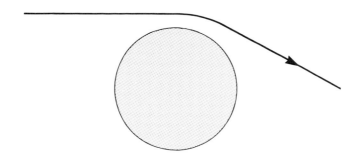

Fig. 2.6 Bending of light. A key prediction of Einstein's General Theory of Relativity is that the path of a ray of light is curved by the presence of matter. The effect is shown here highly exaggerated: light grazing the surface of the sun is deflected by only 1.8 arcseconds.

experimentally. First, the bending of light around the sun, which was first detected in the eclipse expedition of 1919 and was tested much more accurately with radio measurements during the 1970s. Secondly, the phenomenon of the gravitation lens, in which the bending of light around a galaxy amplifies the signal from a much more distant source behind it in the same line of sight. This phenomenon was discovered in 1979 (see Chapter 3 for more detail). General Relativity also predicts the phenomenon of gravitational waves and though, despite several decades of intensive effort, these have not yet been detected directly, they were discovered indirectly in 1975 through the slowing down of the orbital period of a system known as 'the binary pulsar' in which two pulsars, or pulsating radio sources, are in close orbit around each other. Finally, there is the phenomenon of the black hole in which General Relativity predicts that a collapsing massive star would bend light around it so strongly that no signal could escape from it. Such systems have almost certainly been found in several binary X-ray sources in the Milky Way and in the nuclei of many external galaxies. Thus the theory is not just an elegant piece of mathematics. It has survived many experimental tests. Incidentally, black holes would not be of much interest if we did not believe we had found them in the universe. Calculations currently being made in General Relativity on 'worm-holes' connecting different universes, time-travel by moving in and out of such worm-holes, and so on, seem at the moment to be of little importance because there is no prospect of their having any practical application.

It is worth emphasizing that even the most successful physical theories tend to have the same problems of incompleteness that mathematics does. Even those two pillars of modern physics, General Relativity and Quantum Theory, are not complete theories. There are deficiencies in the theories which show that better theories will ultimately be needed. In the case of the General Theory of Relativity, there is no prescription for the large-scale 'topology' of space-time, how things connect up on the large scale. So when we look out to great distances in one direction in the universe, we have no idea how or whether this connects up with, say, the opposite direction on the sky. In some models there is a connection, and if you keep travelling in one direction in the sky, you will eventually end up returning to your starting-point from the opposite direction in the sky. But in other models there is no such connection, and the matter is arbitrary as far as General Relativity is concerned. Thus General Relativity does not give us a complete model of the universe. This seems to me to be a more pressing problem than that of finding a grand unification of General Relativity and Quantum Theory, to which far more effort is being applied. Of course, the solution to the topology problem might come with the solution to the grand unification problem.

In the case of Quantum Theory, the problem has been much more widely discussed: we do not know how to interpret the probabilistic nature of the predictions of the theory. The theory tells us how to calculate the probability that a radioactive atom will decay in a given time, but does not tell us when this will occur. Is the universe at heart probabilistic in nature, as the theory seems to say, or will a deeper and more predictive theory be found? Einstein, for example, refused to accept that the universe was probabilistic in nature, expressing this in the aphorism 'God does not play dice'. One interpretation of the theory says that it is the process of observation that gives reality to the mathematical functions that express these probabilities, so that the universe needs (intelligent?) observers to become real. Is this reasonable? Or do you prefer the rival view that there are many worlds coexisting simultaneously in which all the different possible outcomes are expressed? Again it is clear that, however successful, Quantum Theory is incomplete at present.

Einstein made much of the wonder that mathematical physics is possible and this has been echoed by subsequent writers. But is it really so surprising? Surely it is only a question of having a versatile enough mathematics. We tend to remember only the successes. For every mathematical model which made some impact on physics, there are ten or a hundred which died a swift death. It is very instructive to look at the extraordinarily ingenious mathematical models of the medium which was thought to be necessary to carry light waves, the *luminiferous aether*, during the nineteenth century. Elaborate mechanical models of this aether as an elastic solid were put together to account for the behaviour of light in a medium such as glass, crystal or water. Once the Special Theory of Relativity appeared in 1905, all these models became pointless if not downright ridiculous.

Perhaps the most wonderful application of the General Theory of Relativity is to the study of the universe as a whole. In 1917, a year after publishing his theory, Einstein addressed the question of constructing a model for the universe. He found that there was a problem. As the theory stood he could not construct a stable, static model for the universe. In order to construct such a model for the universe he took the fateful step, which he was later to describe as 'the worst blunder of my life', of postulating a new force, the 'cosmological repulsion', to balance the attractive force of gravity. In constructing this model of the universe, Einstein made the drastic oversimplification that on the large scale the universe was *homogeneous*, that is completely smooth, and *isotropic*, that is looked the same in every direction. There was not a shred of evidence to support these assumptions. On the contrary, there was already good evidence in 1917 that the matter in the universe was very inhomogeneous, gathered into the form of a single gigantic system, the Milky Way Galaxy, surrounded by virtual emptiness, and that the universe had a very definite plane of symmetry defined by the Milky Way, so was far from isotropic. It is remarkable that Einstein's dramatic guess about the simplicity of the universe on the large scale should prove to be so accurate.

Einstein's goal of making a static model for the universe, however, proved to be misguided. In the same year, 1917, the Dutch mathematician Willem de Sitter showed that an expanding universe was also a theoretical possibility within General Relativity. And in 1924 the Russian mathematician Alexandr Friedman discovered the whole set of possible homogeneous, isotropic, expanding universe solutions in

General Relativity. But it was not until 1929 that Edwin Hubble was to show that the universe is in fact expanding. Hubble's discovery and the Big Bang model of the universe to which it gave rise will be the subject of Chapter 4. Although Einstein later repudiated the cosmological repulsion, in recent years it has taken on a new significance within the framework of 'inflationary cosmology', which I will discuss in later chapters. Before we set off down the cosmological trail, though, I give in Chapter 3 an introduction to the new astronomy of the invisible wavelengths.

Thirty years of the new astronomy

ASTRONOMY IS UNUSUAL among the sciences in that there is no real possibility of doing experiments, in the conventional sense of a controlled experiment in a laboratory. Instead we stare out at the cosmos with a bewildering array of telescopes and detectors. One of the striking features of the past thirty years has been the growth of the new astronomy of the invisible wavelengths. In this chapter I review some of the achievements of this new astronomy. The emphasis of this book is on two of the invisible wavebands, the far infrared band, surveyed by the IRAS satellite in 1983, and the microwave band, in which Penzias and Wilson first discovered the cosmic background radiation in 1965 and in which the COBE satellite discovered the 'ripples' in 1992.

However not quite everything we know about the universe derives from studies with telescopes. In the solar system we can study the cosmic environment directly. Some of these studies are a natural extension of terrestial science. In fact there is no sharp distinction between geophysics and astronomy. The earth can be viewed as a part of the solar system, which is in turn a part of our Galaxy and so on.

25

What we learn about the universe from the solar system

In the solar system we can now study individual planets, satellites and comets through space probes. Studies of the surface features of the moon, Mars and Venus, of planetary atmospheres, of models for the interiors of planets and satellites, and of the dust and plasma of the interplanetary medium, all have their parallels in terrestrial studies. Some of these studies resemble terrestrial geography and geology. An enormous amount has been learnt about the history of the solar system and its formation through careful chemical analysis of lunar and meteoritic samples.

These chemical studies also have implications for understanding the evolution of our Milky Way Galaxy and of the universe. For example, studies of long-lived radioactive isotopes give an estimate of the age of the Milky Way in the range 10 to 15 billion years. Obviously the universe cannot be younger than this. Actually since galaxies probably form fairly early in the history of the universe, within the first few hundred million years, the age of the Galaxy and the universe do not differ by much.

Studies of the relative abundances of the elements in the earth and the solar system, as well as in the stars, give clues to the origin of most elements either in normal thermonuclear processes in the centres of stars or in dramatic stellar explosions at the end of a star's life. The light elements deuterium (heavy hydrogen), helium and lithium are especially significant, though, for these were synthesized in the early universe during the first few minutes after the Big Bang. Their abundances relative to hydrogen pin down the average density of ordinary matter in the universe today.

Once we want to step outside the solar system and try to study the universe on a larger scale, we can only make progress through observation with telescopes. There are still some analogies with geology, because we try to unravel the past history of stars, galaxies and the universe through the study of relics and strata of different ages. In general, though, astronomy is a branch of physics. For the universe is full of strange and unfamilar conditions, to which we must extrapolate our terrestrial knowledge. Many fundamental discoveries of physics, from gravitation to thermonuclear fusion have been made by astronomers. This should not blind us, though, to the particularity and uniqueness of the universe.

The opening up of the invisible wavelengths

The past thirty years have been a golden age of astronomy, for we have burst out of the narrow visible waveband to which we were confined for thousands of years by the limitations of our own eyes. Today we are at the end of a dramatic era – the era of the opening up of the invisible wavelengths for astronomy. With each new waveband we have made a new map of the universe, a map containing features which were previously invisible to us. This explosion of knowledge has taken place both with extraordinary new kinds of ground-based telescope and with telescopes launched into space.

The first step towards today's astronomy of all the wavebands came with Newton's realization that the light from the sun was made up of light of many different colours. This visible spectrum was already known, but not understood, as the phenomenon of the rainbow. With Huygens and Young's wave theory of light came the realization that these colours should be identified as light of different wavelengths, or equivalently, different frequencies. Then in 1800 William Herschel discovered that beyond the red end of the spectrum of the sun there were invisible wavelengths which could be associated with radiant heat, the

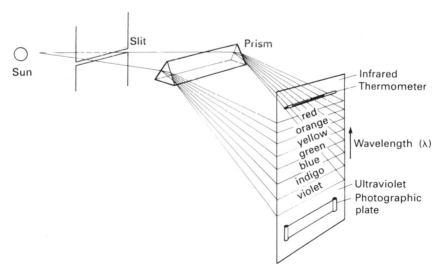

Fig. 3.1 The generation of the visible spectrum by passing light through a prism, with infrared and ultraviolet bands indicated.

infrared wavelengths. Shortly afterwards Johann Ritter found that invisible wavelengths beyond the blue end of the spectrum, the *ultraviolet* wavelengths, could fog a photographic plate. At the end of the nineteenth century came the discovery of X-rays beyond the ultraviolet, and of radio-waves beyond the infrared. We now call this complete range of wavelengths *the electromagnetic spectrum*, following James Clerk Maxwell's demonstration that all forms of light are in fact electromagnetic waves. Astronomers arbitrarily divide the wavelengths into wavebands, which in order of increasing frequency, or of decreasing wavelength, are: radio, microwave, submillimetre, far infrared, near infrared, visible, ultraviolet, extreme ultraviolet, soft X-ray, hard X-ray and gamma-ray. Here 'near', 'far', 'extreme' refer to how different the

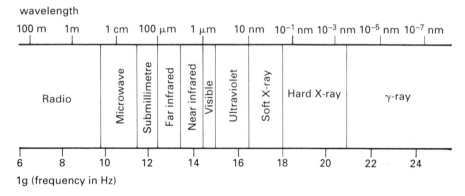

Fig. 3.2 The wavebands of the electromagnetic spectrum, with their frequencies and wavelengths.

wavelengths are from those of the visible band, and 'soft' and 'hard' refer to lower and higher energies, that is lower and higher frequencies. This division into wavebands is based primarily on detection techniques and has no fundamental meaning. The visible waveband, stretching from 0.4 to 0.7 microns in wavelength (a micron, or micrometre, is one-millionth of a metre) is surprisingly narrow, compared with the whole range of wavelength spanned from the metre wavelengths of the radio band to the one ten thousand millionth of a micron wavelengths of the gamma-ray band.

Only certain of these forms of radiation manage to penetrate earth's atmosphere to reach the ground: the radio, the microwave, the optical and some of the ultraviolet wavelengths. Some submillimetre and near infrared wavelengths become accessible to telescopes on mountain-tops. Water vapour, carbon dioxide, oxygen and other gases of earth's atmosphere absorb most of the infrared wavelengths. Ozone and other atmospheric gases absorb the ultraviolet, X-ray and gamma-ray wavelengths. For these wavelengths some glimpses of the universe can be achieved with airborne or rocket-borne telescopes, but for real progress we have to put telescopes into orbit around the earth on satellites.

After Herschel's discovery of infrared radiation from the sun, the next astronomical object to be detected at infrared wavelengths was the moon, detected in 1856 by the Astronomer Royal for Scotland, Pazzi Smyth, during an expedition to the island of Tenerife. During the 1920s near infrared measurements began to be made of some planets and stars, and ultraviolet measurements of stars also become routine. These were important in helping to determine the temperatures of these bodies more accurately. But the opening up of the electromagnetic wavebands really got going with the rapid growth of radio-astronomy in the late 1950s and early 1960s. The American radio-engineer, Karl Jansky, had discovered radio emission from the Milky Way in 1934 while working at Bell Telephone Laboratories on causes of noise in transatlantic telephone lines. Judging that this discovery did not have much practical application, the Bell Telephone company did not permit Jansky to follow up his discovery, assigning him to more conventional telecommunication activities. Few astronomers took much interest in Jansky's discovery over the next decade. Radio astronomy was kept alive only by the work of the American amateur astronomer Grote Reber, who painstakingly mapped the Milky Way with a radio-telescope he had constructed in his back garden. It took the war-time discovery by the British engineer John Hey and his radar research group of radio emission from the sun, and of the first strong point-like source of cosmic radio waves, to stimulate others who had worked in radar to develop radio-astronomy after the war.

The thirty years from 1960 to 1990 were the period when human beings learnt to see the universe in new kinds of light. In the late 1950s and 1960s the radio band was opened up by the pioneering surveys of the Cambridge group, led by Martin Ryle, and of the Australian group at Parkes, led by John Bolton. In 1965 came the discovery of the microwave background radiation by Arno Penzias and Robert Wilson (the subject of Chapter 8), in 1970 and then their discovery of microwave radiation from the carbon monoxide molecule, the first of many organic molecules to be discovered in interstellar space over the following decade. In 1970 too came the launch of the pioneering Uhuru satellite, with the first survey of the sky at X-ray wavelengths, followed in 1978 by the Einstein X-ray satellite observatory, both projects led by Riccardo Giacconi. The ultraviolet band was probed first with the Copernicus satellite in 1972 and then with the International Ultraviolet Explorer satellite, launched in 1978 and still working today, fifteen years later. Finally, the sky was surveyed in gamma-rays with the Cos-B satellite launched in 1975, and in the far infrared with the Infrared Astronomical Satellite (IRAS) in 1983.

Are we at the end of the golden age?

1990 saw the launch of ROSAT, opening up the extreme ultraviolet and soft (lower energy) X-ray wavebands, and discovering tens of thousands of X-ray sources in its all-sky survey. 1990 also saw the launch of the Hubble Space Telescope, which in spite of its optical defects is discovering exciting things in the far ultraviolet. Although the Hubble Space Telescope will not reach its full design performance for optical imaging until after the repair mission, due for launch on the Shuttle in late 1993, it has already achieved some impressive successes, including a measurement of the distance of a key galaxy on the cosmological distance ladder (see Chapter 4). The Gamma-Ray Observatory, now renamed the Compton Observatory, was launched in 1991 after a long wait in the Shuttle queue and has detected scores of active galaxies as well as many so far unidentified sources. That leaves only the submillimetre band – 100 microns to 1 millimetre – relatively unexplored, though ground-based work in the 800 and 350 micron windows and observations from the Kuiper Airborne Observatory at 100 – 300 microns are nibbling away at this waveband too. And the Cosmic Background Explorer (COBE), launched in 1989, has already started to give some fascinating images of our Milky Way Galaxy at these wavelengths, as well as discovering the primordial 'ripples' from which the structure we see in the universe today, galaxies, clusters of galaxies, and even larger structures, developed. It is the meaning of these ripples which I have made the main theme of this book.

The fact that there have now been surveys of the sky in most of the electromagnetic wavebands does not, of course, exhaust the possibilities for new astronomical discoveries in any of them. Otherwise astronomers would hardly be building still larger ground-based telescopes in the optical band, which has been used for astronomy for thousands of years. Important space astronomy missions of the 1990s include the European Space Agency's Infrared Space Observatory (ISO) and XMM missions, and NASA's Advanced X-ray Astronomy Facility (AXAF), which is due to follow the Hubble Space Telescope and the Compton Observatory as the third of the Agency's 'Great Observatory' series. But we are now at the point where future missions are likely to be increasingly complex and specialized and may not have quite the same frontier feel as the missions of the past. We appear to be approaching the end of a golden age, when we had our first glimpse of the universe in the invisible wavelengths.

Our place in the cosmic landscape

What have we learnt from these 30 years of the new astronomy? To sum it up in one theme, what we have learnt is the intimate relationship between human existence and the universe – our history and our destiny are bound up with the stars to a degree undreamt of by the astrologers of the Middle Ages. (Of course there is no clue from the stars to our personal destiny and the gullible perusers of horoscopes are doomed to be disappointed.) Is the universe tuned to permit our existence to such a remarkable degree that some special explanation, the 'anthropic principle', is required? This has been proposed by a number of scientists, for example Brandon Carter, Martin Rees, Bernard Carr, John Barrow and Frank Tipler, a speculative argument I will return to later. First I want to summarize some of the variations on this theme of our place in the universe.

At the core of modern astrophysics is the life-cycle of stars: their formation in dense clouds of molecular gas and dust; their evolution as stable stars through the fusion in their cores of successive nuclear fuels, first hydrogen to helium, then helium to carbon, nitrogen and oxygen, and so on through neon, magnesium and silicon to iron; and their convulsive death throes with their outer layers being thrown off with greater or lesser violence, and their cores being compressed to the outlandish white dwarf, neutron star or black hole states. The elements of

which the earth and our bodies are made were first created in the fur-
naces at the centres of stars or in the dramatic deaths of massive stars
as supernovae.

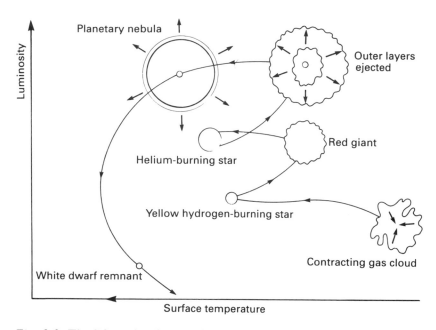

*Fig. 3.3 The life-cycle of a star like the sun. The star forms by the conden-
sation of a cloud of interstellar gas and dust. It spends most of its life fusing
hydrogen to helium. As hydrogen is exhausted at the centre, the star expands
to become a red giant. The core then collapses and heats up until helium starts
to fuse to carbon. As the helium is exhausted the star becomes a red giant
again, ejecting its outer layers to form a planetary nebula. The exposed core
cools off to form a white dwarf remnant. Stars of mass of more than ten times
that of the sun end their lives in a much more dramatic supernova explosion,
with the core collapsing to form a neutron star or black hole.*

The first phase, fusion of hydrogen to helium, is the longest last-
ing and is common to all stars. The sun and most of the stars we see
in the night sky are in this phase. When hydrogen is exhausted in the
core of the star, this core contracts and heats up until the next fuel,
helium, ignites. Meanwhile, the star brightens and the outer layers of
the star expand enormously and cool, until the star becomes a 'red

giant' star. During the red giant phase stars lose mass profusely from their surfaces in violent 'stellar winds'. A star like the sun will eventually throw off the whole of its outer layers in a 'planetary nebula' event, exposing the compressed hot core as a 'white dwarf' remnant, which slowly cools down. More massive stars proceed through a series of nuclear fuels up to iron. Once the core is composed of iron, nuclear fusion ceases, the core suddenly collapses inwards to form an even more highly compressed 'neutron star' or 'black hole', and the outer layers of the star are ejected in a spectacular 'supernova' explosion.

The new astronomy has shed light particularly on two phases of this life-cycle of stars. First, infrared astronomy has drawn back the veil of dust which shrouds the birth of stars from view, with many thousands of stars in the process of forming having been identified by the IRAS satellite. And secondly, the death throes of stars have been probed in almost all the wavebands, firstly with the discovery of pulsars, which are rapidly rotating neutron stars, at radio wavelengths; then with studies of the phase of mass-loss and planetary nebula formation of red giants at infrared and wavelengths; and with X-ray astronomy revealing how the dead remnants – white dwarfs, neutron stars or black holes – are lit up in binary systems (two stars orbiting around each other) by mass transfer from their companions.

All that was known of the material between the stars prior to the new astronomy was that visible light was dimmed by dust. Radio astronomy showed us the all-pervasive atomic hydrogen gas between the stars, the microwave band showed the dense clouds of molecular gas where new stars are formed and in the infrared we saw the emission from dust grains. These studies allowed us to identify the different constituents of these grains – silicates, carbon, ices, organic molecules. From these grains, in which most of the heavier elements reside after they are driven out from their place of formation in the stars, the earth was assembled. The carbon of our bodies was first in the interior of a star, then in a fierce wind blowing off the star, then in a minute dust grain swirled through the space between the stars, before finding itself part of the solar 'nebula', from which the sun, planets and earth itself formed.

That process of the formation of the sun and planets is still far from being completely understood. But the study at infrared and microwave wavelengths of stars in the process of formation has generated a

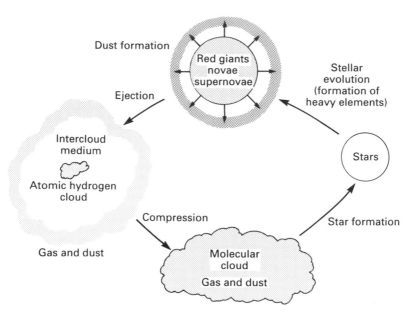

Fig. 3.4 The cycle of gas and dust in the interstellar medium. Heavy elements formed in the interior of stars are ejected in stellar winds, planetary nebulae or supernova explosions, and condense into small grains of interstellar dust. Clouds of gas and dust aggregate together and, when the density is high enough, form molecular hydrogen. New stars then form when the dense molecular clouds undergo compression.

picture which may be universal for single stars. At the centre of the rotating cloud of gas and dust, a 'protostar' begins to condense. The outer parts of the cloud settle into a doughnut-shaped disc, from which gas flows inwards towards the protostar. The centre of the protostar becomes hot enough for fusion of hydrogen to begin. A wind sweeps out from the star and is focused into two opposite cones by the disc. From the disc itself, the planets will in time form.

On the periphery of the solar system resides a cloud of comets, primitive aggregates of dust, rocks and ices. From time to time one of them is perturbed by a chance gust of gravity, the effect of some passing star, and it plunges in towards the inner reaches of the solar system where the ices melt and fluoresce and the smaller dust grains are driven out as a tail of debris. After many passages nothing remains but the rocky core of the comet, plunging through the solar system as a dark asteroid. Such an object was the asteroid IRAS 1983TB, now known as Phaethon, which is the nucleus of a long dead comet whose debris is still visible to us once a year in the December Geminid meteor stream. The surface of the moon is scarred with the ancient impacts of these huge rocks. Earth too is endlessly bombarded with cometary debris. In these violent impacts we are witnessing the vestiges of the processes by which the planets were assembled from asteroid-sized 'planetesimals' aeons ago.

The new astronomy has transformed our picture of galaxies, both in mapping the gaseous and dusty discs of spiral galaxies like our own, and in showing us the active nuclei in the centres of some galaxies. The vast double lobes of radio-galaxies and quasars are now seen to be powered by beams of particles moving close to the speed of light generated in massive black holes. More recently the Infrared Astronomical Satellite, IRAS, has shown that similar enormous powers can be produced through star formation on an extraordinary scale, probably driven by interactions and mergers between galaxies which have passed too close to each other.

The stars and nebulae of the night sky

In all these aspects we see our existence, our history and our future bound up with the universe of stars and galaxies. Very little of this story was known before the birth of the new astronomy. Does this mean that astronomy has become an esoteric and invisible science? I believe not, because almost all of the insights of the new astronomy can be linked to familiar objects of the night sky or to objects visible with binoculars which have been known for centuries.

When we stare at the belt of Orion, the Hunter, we are looking at the nearest dense molecular cloud where new massive stars are in the process of being born. With Mira, the 'Wonderful', we are seeing the penultimate stages of a star like the sun as it blows off its outer layers and prepares for its final convulsion as a planetary nebula. Orbiting Sirius, the 'leader of the host of heaven', is its white dwarf companion, last relic of a star like the sun when it completes its ten billion year life-span. Betelgeuse, the Armpit of the Giant, is a massive star on the brink of exploding as a supernova like that of 1987 in the Large Magellanic Cloud, or that of 1054 AD, recorded by the Chinese astronomers and perhaps by the Navajo Indians of Northern Arizona, and visible today with binoculars as the Crab Nebula. In Algol, the Demon Star, the eye of the Medusa held aloft by Perseus, we see the prototype of close interacting binary stars where mass is transferred from one star to another, often with dramatic results at X-ray wavelengths if one of the stars is already a dead, compact remnant like a white dwarf, a neutron star or a black hole, for then the infalling gas is heated to temperatures of tens of millions of degrees and radiates X-rays. Vega, high overhead on a northern summer night, was found by IRAS to be surrounded by a disc of dust particles, perhaps a planetary system in the making. Looking up at the Milky Way on a clear night, the dark outline of the Coalsack shows strikingly the existence of matter between the stars, a vast cloud of dust and gas. The Magellanic Clouds, visible to the naked eye in the southern hemisphere, are our nearest neighbour galaxies. They are doomed to spiral in towards the Milky Way and be swallowed up by our Galaxy before many hundreds of millions of years have passed. In Andromeda, the Nebula visible to the naked eye is our Galaxy's dominant partner in the Local Group of the thirty or so nearest galaxies. With binoculars many of the galaxies that make up the rich cluster in the constellation of Virgo, first seen by William Herschel 200 years ago, can be seen, amongst them Messier 87, prototype of the powerful radio-emitting radio galaxies, with their twin lobes of radio emission extending millions of light years on either side of the central nucleus of the galaxy. In the constellation of the Great Bear, so conveniently defined by the Plough, or Big Dipper, lies Messier 82, also visible with binoculars, the nearest of the starburst galaxies, undergoing a strong interaction with its beautiful spiral companion, Messier 81.

These objects, many of which have been written about by writers of all ages, form the bridge between the modern science of astronomy, with its advanced technology telescopes on mountain-tops and in space, and the rest of human culture. Of course there are not many people today, even, I have to admit, in the ranks of professional astronomers in this age of computer controlled telescopes, who are familiar with the night sky or can identify more than a few of the most obvious constellations. Yet it was once very different. Two or three thousand years ago, people had a thorough knowledge of the night sky and its motions. This is very clear from the many astronomical references in ancient literature, whether it be the Chinese lyric poets or the classical Greek and Roman writers. For Dante, astronomy was central to his vision of the world. The writings of Shakespeare and his contemporaries are full of astronomical allusions. But it is unusual to find astronomical references in modern writing.

This apparent indifference to the starry night does not just date from the moment when everyone uprooted and made for the smoke-shrouded cities. Surely the great Romantics, Goethe, Byron, *they* must have known the night sky? Not a bit of it. They talk eloquently of 'the stars' but it's hard to find much evidence that they knew the name of a single one. Since the sixteenth century, writers who have been familiar with the night sky – for example Milton, Tennyson, Thomas Hardy, Italo Calvino – have been the exception rather than the rule. But of course the stars, the universe, have remained a powerful image in all centuries.

Perhaps people stopped looking at the night sky because after Copernicus and Galileo everything seemed to be explained. The erratic motions of the planets along the zodiac held no more mystery, no longer was it rational to imagine that comets were terrible omens. Human destiny was not tied to the night sky and so we gradually stopped looking at it. But the discoveries of modern astronomy show that both our history and our destiny *are* bound up with the stars. The astronomers of the present age, with their superb telescopes on the tops of mountains or orbiting the earth on spacecraft, have given us images of the stars and galaxies unimaginable to the ancients. And with these images has come a new insight into the nature of the stars and the universe.

At least two of the phenomena of the new astronomy are, as it were, out of the blue and cannot be related to any object of the night sky that is visible with naked eye or binoculars. The first of these phenomena is the quasar, short for quasi-stellar radio-source, for the brightest quasar, 3C273, lies at 12th magnitude, 300 times fainter than the faintest star visible to the naked eye. I will return to these in Chapter 11. The second is the phenomenon of the microwave background radiation, interpreted as the relic of the fireball phase of the Hot Big Bang, on which so much of our present cosmological story is pinned. This can be experienced gratuitously by looking at the random dots on your television screen after broadcasting has ceased.   Part of this 'noise' is due to the microwave background radiation. This radiation is also in a sense related to the *darkness* of the night sky, which in a supposedly infinite universe of stars perplexed astronomers and philosophers for centuries (the so-called Olbers Paradox) until Edgar Allan Poe explained the darkness as due to the finite age of the universe. The light from the most distant stars has not had time to reach us yet. But as I like to point out, the sky is pretty bright in the microwave band, being about as bright in energy terms as the Milky Way is in the visible band. The microwave background radiation will be the subject of Chapter 8.

Is our relationship with the universe so strong that we have to regard the physical parameters of the universe as having been fine-tuned so that we could exist? This is the basis of the *anthropic principle*, a fascinating, metaphysical speculation that is impossible to test. The argument is that, for example, if gravity were a bit weaker or a bit stronger, stars would not exist, so neither would we. This can be extended to almost all physical laws and chemical properties. But since we have not been particularly successful in demonstrating how galaxies or stars form in this universe, nor how life arose on this earth, it seems premature to be predicting what might or might not have happened if the universe had been different. This is the universe we have and what seems important is to understand and admire it.

Hubble's law, the cosmological distance ladder and the Big Bang

IN 1929 THE American astronomer Edwin Hubble discovered the expansion of the universe. He found that the spectral lines of distant galaxies are shifted towards the longer wavelength, red end of the spectrum. If we interpret this as a 'Doppler' shift, this means that the galaxies are moving away from us. The Doppler shift is more familiar in the context of sound waves. As a whistling train approaches us, the pitch or frequency of the whistle is raised, as it recedes from us the pitch is lowered. The effect was discovered by the Austrian physicist Christian Doppler in 1842, and he demonstrated it with a band of musicians in a moving railway carriage. For light, the relative shift in wavelength, or *redshift*, is simply the ratio of the recession velocity to the speed of light.

Hubble also found that the redshift, and hence the recession velocity, increases with distance from us. This is true whichever direction the galaxy is in. The simplest explanation is that the whole universe is expanding. At first sight it might seem that if the galaxies are moving away from us in every direction, then we are at the centre of the universe. This is an illusion, however, as an observer sitting on any other galaxy would see the same thing, themselves at the centre and the other galaxies moving away from them in every direction.

Fig. 4.1 Edwin Hubble (1989–53).

The importance of Hubble's discovery of the expansion of the universe can hardly be overstated. It is a milestone in human history comparable to Copernicus's proof that the sun is the centre of the solar system. In this chapter I discuss the evidence that we live in an expanding universe and the controversies that continue to rage about the size and age of the universe. The painstaking sequence of methods of measuring cosmic distances, reaching out from the solar system to the most distant galaxies, which I have called the *cosmological distance ladder*, underpins everything else that we try to do in cosmology. I also describe the Big Bang model of the early universe and the current view that very early on the universe went through a phase of spectacular 'inflation'.

Hubble's Law

On its own, the Hubble redshift distance law is not conclusive proof of expansion, for we might imagine that some other physical effect could cause a redshift. However, other types of explanation did not stand the test of time, whereas the expanding universe picture has received many independent confirmations. One factor in the immediate acceptance of Hubble's discovery as evidence for an expanding universe was the fact that expanding universe models had been developed from Einstein's General Theory of Relativity by de Sitter, Einstein, Eddington, Lemaître and Friedman in the period 1917–24. But there was an immediate problem. The time-scale for the expansion of the universe implied by Hubble's measurements, 2 billion years, was soon found to be inconsistent with the age of the earth. In the simplest General

Fig. 4.2 M31, a nearby spiral galaxy similar to our Milky Way galaxy. M31 and our Galaxy are the dominant members of the Local Group of galaxies.

Relativistic cosmological models, the age of the universe always turns out to be less than the expansion time, because the self-gravitation of the matter in the universe slows the expansion down. To believe in the expanding universe it was necessary to believe that one or other of these time-scales had been wrongly measured. This did turn out to be true, but for two decades this discrepancy posed a severe headache for cosmologists. It was to explain the discrepancy that Hermann Bondi, Tommy Gold and Fred Hoyle invented the Steady State Cosmology in 1948. Today, improved distance estimates put the expansion time-scale at 10–20 billion years, compared to the earth's age of 4.5 billion years and the age of our Milky Way galaxy of 10–15 billion years.

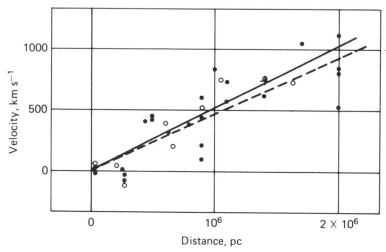

Fig. 4.3(a) Hubble's 1929 velocity-distance law. The most distant galaxies belong to the Virgo cluster.

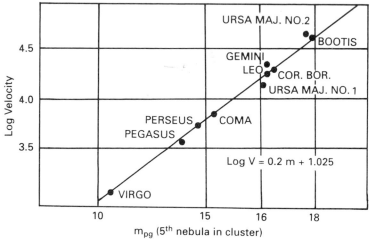

(b) The velocity distance law published by Hubble and Humason in 1931.

The controversy about the distance scale can be encapsulated in what is known as the Hubble constant, H_o (pronounced 'H-nought' or 'H-zero'), which is the coefficient of proportionality in the Hubble law. For redshifts which are a small fraction of 100% (say less than 10%), the redshift can be translated into a velocity by

$$\text{redshift} = \text{velocity of galaxy} \ / \ \text{speed of light}$$

(This is essentially a Newtonian equation and it breaks down when the recession velocity approaches the speed of light. However, almost all the measurements I will be describing, on which the Hubble law is based, refer to galaxies with redshifts of less than 0.1, i.e. a 10% increase in wavelength. In Chapter 12, I will be talking about quasars and other very distant galaxies, for which redshifts approaching 5 (i.e. all wavelengths have been increased by 500%) and for these the concept of a recession velocity is not very useful.)

The Hubble law

redshift *proportional to* distance

can therefore be written

$$\text{velocity} = H_0 \times \text{distance,}$$

where H_0 is the coefficient of proportionality.

Astronomers measure velocities in kilometres per second and distances in megaparsecs (1 megaparsec = 1 million parsecs, 1 parsec = 3.2615 light years), so the units of H_0 are kilometres per second per megaparsec (or km per sec per Mpc). The inverse of the Hubble constant then has the dimensions of a time, the expansion of the universe (the time-scale for the universe to double in size, expanding at the present rate). Hubble's 1929 value for H_0 was 500 km/s/Mpc, and since there are 3.09×10^{19} kilometres in a megaparsec, and 3.16×10^7 seconds in a year, this corresponds to a time-scale of

$$3.09 \times 10^{19} \, / \, (500 \times 3.16 \times 10^7) = 2 \text{ billion years}$$

The data on which Hubble based his law (Fig 4.3a) were poor. The most distant galaxies in his sample lay in the Virgo cluster and had velocities of only 1000 kilometres per second. Today we would see such galaxies as part of our own locality, not necessarily representative of the universe as a whole. However, working with his assistant Milton Humason, Hubble was soon able to extend his velocity-distance relation to much greater distances and in 1931 they published a much more impressive relation reaching to velocities of 20,000 km/s, 7% of the speed of light (Fig 4.3b). Over the next twenty years Hubble, Humason and Nick Mayall gathered redshifts for 850 galaxies, reaching out to velocities of 100,000 km/s, or ⅓rd of the speed of light.

There were a number of mistakes in Hubble's distance measurements, which were gradually corrected over subsequent decades. In 1946 the Swedish astronomer Kurt Lundmark pointed out that there was a discrepancy between the distance of the Andromeda galaxy M31 deduced by Hubble, and the larger distance estimates found by comparing novae or globular clusters in M31 with those in our own Galaxy. To estimate the distance of M31 and other nearby galaxies, Hubble had used a particular type of variable star called a Cepheid variable, named after the star Delta Cephei, in which the period of variation is correlated with the star's luminosity. This correlation had been found by Henrietta Leavitt while studying variable stars in the Magellanic Clouds in 1912. By comparing Cepheids in an external galaxy with similar ones in star clusters of known distance in our own Galaxy, the distance of the external galaxy can be found. In 1952 the American astronomer Walter Baade found there are in fact two types of Cepheid which satisfy very different period-luminosity relations. The Cepheids used by Hubble in our own Galaxy were a factor two lower in luminosity than those in the external galaxies he was studying. Baade deduced a Hubble constant of 250 km/s/Mpc.

For more distant galaxies, Hubble had used the brightness of the brightest stars in the galaxy to measure distance, but in 1956 Allan Sandage found that the objects studied by Hubble were not individual stars but much more luminous clouds of hot gas. In a 1956 paper with Humason and Mayall summarizing the results of Hubble's programme (Hubble had died of a heart attack in 1953), Sandage added a footnote revising the Hubble constant to 180 km/s/Mpc. However, Sandage soon found that the error was more serious than he had at first thought and in 1958 he concluded that the Hubble constant was 75 km/s/Mpc. In retrospect we can see that this was the first reasonably accurate estimate of the Hubble constant. Between 1929 and 1958, estimates of the Hubble constant had shrunk by a factor of 7. The estimates of the size of the universe and of its expansion time-scale had increased by the same factor. Considering the magnitude of the errors he was making in his distance estimates, and the small volume of the universe he was surveying, Hubble was perhaps lucky to discover his famous law with his 1929 data. Since 1958 almost all measurements of H_o have been in the range $50-100$ km/s/Mpc, corresponding to expansion time-scales of $10-20$ billion years.

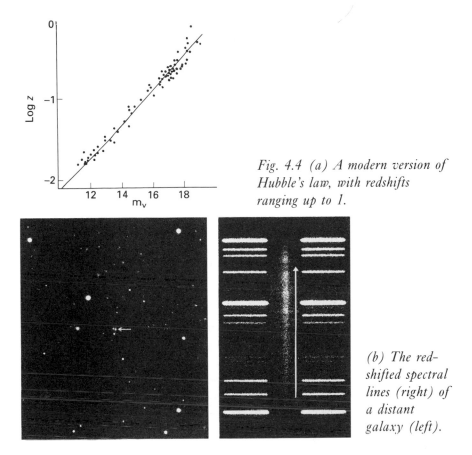

Fig. 4.4 (a) A modern version of
Hubble's law, with redshifts
ranging up to 1.

(b) The red-
shifted spectral
lines (right) of
a distant
galaxy (left).

The Shapley – Curtis debate about the spiral nebulae

In this century, there have been three great controversies about the cosmological distance scale. The first, which raged for the first twenty
years of this century, and had its roots in work of the previous hundred
years by William Herschel and Lord Rosse, was about the distance of
the spiral nebulae. Many of the nebulae, or fuzzy looking objects, first
catalogued by Messier and Herschel in the eighteenth century, had
turned out from spectroscopic studies to be relatively nearby clouds of
gas, like the Orion Nebula. However the nature of those nebulae which
Lord Rosse, in the middle of the nineteenth century, had shown to be
spiral in shape remained controversial. In 1900 most thought that the
nebulae were part of the Milky Way system. The controversy culmi-

nated in a famous debate between Harlow Shapley and Heber Curtis at a meeting of the American Association for the Advancement of Science in 1920. Shapley had been measuring the size of the Milky Way and thought he had shown it was large enough to encompass spiral nebulae like the Andromeda Nebula, whose distance was not very accurately known at the time. Curtis, on the other hand, was convinced that the spiral nebulae were external systems similar to the Milky Way and he marshalled a number of arguments to support this. The debate was inconclusive and the issue was not settled, in Curtis's favour, until Edwin Hubble's work on Cepheid variable stars in galaxies in the mid-1920s showed that these galaxies lay far beyond the confines of the Milky Way. It was this work which led on to Hubble's famous discovery of the expansion of the universe.

The nature of quasar redshifts

The second controversy was about the nature of the redshift of quasars and this reached its peak in the early 1970s. The extraordinary properties of quasars, especially their phenomenal luminosities, equivalent to hundreds of times the total output of our Galaxy, shown by the rapid variability in the brightness to be coming from a region not much larger than the solar system, suggested to some astronomers that the redshifts of quasars might not be anything to do with the expansion of the universe. Perhaps instead the redshifts might be due to strong gravitational fields in the quasars, or to huge velocities of ejection from some local explosion. Advocates of a 'local' origin for quasars included James Terrell, Halton Arp, Geoff Burbidge and Fred Hoyle. At that time I noticed that all the arguments in favour of cosmological distances for quasars applied only to one sub-class of quasar. It was therefore possible that, as with the earlier controversy about the nebulae, there were two entirely distinct classes of object, one relatively local and one truly cosmological. I remember being quite disappointed that this reasonable argument made little impression on the protagonists of the controversy, who seemed to know from some deep inner conviction that their hypothesis must prevail. As it happens, the supporters of the cosmological view were rather quickly proved to be correct, so this controversy was short-lived. In fact the 'local' theory for quasars never did attract many supporters. The most convincing evidence for the cosmological nature of quasar redshifts was careful work by Alan Stockton, who found that the low redshift quasars were often surrounded by groups of galaxies, whose redshifts turned out to be the same as that of the quasar. Since the redshifts of the galaxies could be presumed to be cosmological, it was reasonable to suppose that the redshifts of the quasars were too.

Modern controversy about the Hubble constant and the size of the universe

The third controversy about the distance scale, which is still raging in modified version today, surfaced in a particularly sharp form at a 1976 conference in Paris on 'The Redshift and the Expansion of the Universe'. There Gerard de Vaucouleurs on the one hand, and Allan Sandage and Gustav Tammann on the other, arrived at estimates of the size of the universe, as measured by the Hubble constant, differing from each other by a factor of two. In one of the discussion sessions at this conference I asked the protagonists what was the range outside which they could not imagine the Hubble constant lying. The ranges they gave did not even overlap. Given that they were studying more or less the same galaxies with rather similar methods, often using the same observational material, I found this incredible. I decided to try to

Fig. 4.5 (a) Allan Sandage,   *(b) Gerard de Vaucouleurs.*

get to the bottom of this paradox and spent the next six years writing my book *The Cosmological Distance Ladder*. I think that I did at least partly succeed in explaining the divergence between the two sides.

Allan Sandage had been hired at the Mount Wilson observatory in 1952, shortly before the death of Hubble. He was given the task of continuing Hubble's cosmological programme and was allocated significant amounts of time on the Mount Palomar 5-metre telescope to do this. He saw the core of the distance scale to be the Cepheids and set out to place the Cepheid method on an impregnable basis, studying them first in the nearby galaxies of the Local Group of galaxies and then in more distant groups. In 1963 the Swiss astronomer Gustav Tammann came to work with Sandage. Between 1968 and 1975 they published a series of papers building a ladder of distance out to the largest cosmological scales. First, they determined distances to several nearby galaxies which, with our own Galaxy, form part of the Local Group of galaxies, and to the galaxy NGC2403 in the nearby M81 group of galaxies, using Cepheids. Then they used galaxies of these two groups to calibrate the next step up the ladder, the luminosity and size of the brightest clouds of ionized gas in galaxies, called 'HII regions' because HII denotes ionized hydrogen. By studying how the size and luminosity of the brightest HII regions in a galaxy were related to the total luminosity of the galaxy, they could estimate the distance of any other galaxy in which HII regions could be detected. Finally, with the galaxies whose distances were known from the HII region method, they calibrated a third distance indicator, the 'luminosity class' of spiral galaxies. The Canadian astronomer Sidney van den Bergh had found in 1960 that the luminosity of spiral galaxies was related to the clarity and contrast of their spiral arms, which he classified into 'luminosity classes'. Thus by classifying the luminosity class of a spiral galaxy, its distance could be estimated. In 1975 Sandage and Tammann used this method to reach out to large distances and give them the Hubble constant free of all local motions, which they found to be 55 km/s/Mpc, with an uncertainty which they claimed was ± 6 km/s/Mpc.

I first met Allan Sandage in 1966 during a visit to Pasadena. I was staying with Michael and Margaret Penston, who were working at Caltech and whom I had first met when we were fellow students a few years previously. Michael had met Sandage and said that he was very approachable, so I rang him and asked to see him. He invited me over to his office in Santa Barbara Street (the headquarters of the Mount Wilson and Palomar Observatories). He was very charming and avuncular and took the trouble to explain his and Tammann's programme to determine the Hubble constant. He showed me the photographic plates he was working on and his 'blink' comparator which allowed two plates taken at different times to be compared, so that variable stars like Cepheids could be identified. He insisted on making me try it out to see if I could find the Cepheids on the plates. It was not easy! Already at that time it was clear to him that a value of around 50 km/s/Mpc was going to emerge, which would solve the problem of the apparent discrepancy between the long ages for globular clusters (18 billion years was his estimate at that time) and the age of the universe (no more than 13 billion years if his 1958 value of the Hubble constant, 75 km/s/Mpc, was correct). With a Hubble constant of 55 km/s/Mpc, the age of the universe could be as long as 18 billion years, long enough to accommodate the globular clusters if they were formed early in the history of the universe. Some years later, stellar evolution theorists arrived at lower estimates of the ages of globular clusters, so this problem was eased.

At the General Assembly of the International Astronomical Union in Grenoble in July 1976, the French astronomer Gerard de Vaucouleurs was billed to give a lecture on the distance scale. He stunned those of us present with a detailed and meticulous assault on Sandage and Tammann's work, querying in his urbane and sarcastic style almost every step they had taken. As Sandage and Tammann were not at the meeting, I felt that someone should say something on their behalf and I commented in the discussion session following Gerard's talk that it was easy enough to criticize their programme, but that it would require an immense amount of work to put a rival distance scale in place. It was in fact to be several years before Gerard's detailed critique and his rival distance scale programme were completed and published. Later in 1976, at the meeting in Paris on 'The Redshift and the Expansion of the Universe' which I have already mentioned, Gerard de Vaucouleurs and Gustav Tammann presented their conflicting views of the cosmological distance scale. It was also at this meeting that the proponents of the view that the redshifts of quasars were not cosmological made their strongest pitch, though in my view the issue had already been settled by Alan Stockton's discovery of many examples of quasars in groups of galaxies at the same redshift as the quasar.

Other criticisms of the Sandage and Tammann distance scale were surfacing. In the same year, 1976, the young Canadian astronomer Barry Madore argued that Sandage and Tammann had overestimated the distance to the galaxy NGC2403 in the M81 group. He believed they had underestimated the effects of interstellar dust in the galaxy. Another young astronomer, the Australian David Hanes, used globular clusters in galaxies to argue in 1977 that the Virgo cluster was not as distant as Sandage and Tammann claimed. In 1979 Richard Kennicutt carried out a detailed study of the HII region method for estimating galaxies' distances and cast considerable doubt on the efficacy of the method. I remember talking to Sandage around this time and finding that he had strong feelings about being attacked in this way by young astronomers, often carrying out their first piece of research.

In 1977 a new distance method was discovered by the American Brent Tully and the Frenchman Richard Fisher, which became known as the Tully–Fisher method. It was based on a correlation they had found between the characteristic rotation velocity of the disc of spiral galaxies and the optical luminosity of the galaxies. A galaxy's rotation can be studied by measuring the profile of a spectral line, for example in the optical region of the spectrum. Tully and Fisher in fact used a spectral line of neutral hydrogen which is emitted in the radio band, at a wavelength of 21 cm. Provided the galaxy is not face on to our line of sight, we will see, superposed on the Hubble recession of the galaxy, one side of the galaxy coming towards us, and the other side moving away. Although the speeds involved are small compared to the recession speed, we see part of the line emission from the galaxy slightly redshifted relative to the average wavelength and part blue-shifted, and this allows us to measure the rotation speed. The faster the galaxies rotated, the more massive and luminous they were, and so distances could be estimated from measurements of the rotation velocity and the apparent brightness of the galaxy. Tully and Fisher's first application of the method suggested a shorter distance scale, and hence a higher Hubble constant, than that found by Sandage and Tammann. The method was siezed upon by de Vaucouleurs and also by Sandage and Tammann. Characteristically each reached radically different conclusions from the method, de Vaucouleurs finding the Hubble constant was 100 km/s/Mpc, while Sandage and Tammann found that the same method gave a value 50 for the Hubble constant.

Gerard de Vaucouleurs' distance scale was, like Sandage and Tammann's, based on three levels of distance estimator, which he characterized as primary, secondary and tertiary. Whereas Sandage and Tammann had used only one distance estimator at each level (Cepheids, HII regions and spiral luminosity classes), de Vaucouleurs, working with a number of collaborators, especially his wife Annette, used as many methods as he could lay his hands on. Some of these methods appeared for the first time in his papers and had not been explored in much detail. While some of Gerard's criticisms of Sandage and Tammann were justified, I found he had introduced a few circular and inconsistent arguments into his own distance scale. Much of the disagreement between the two scales turned out to be in their estimates of distances to the very nearest galaxies of the Local Group. They also disagreed about the amount of absorption of light by interstellar dust in our own Galaxy. In trying to unravel the inconsistencies in the two scales, I rather naturally ended up with a distance scale intermediate between the two of them, and a best estimate of the Hubble constant of 67 km/s/Mpc, with an uncertainty of ± 15 km/s/Mpc. The main problem was not really the actual estimates of the Hubble constant which the two groups arrived at, both of which were possible values, but the overstated accuracy which they had quoted.

In 1982 Sandage and Tammann came up with an entirely new distance scale. They abandoned the HII region and spiral luminosity class methods as too unreliable and instead used two new methods, the brightest red stars in galaxies and Type I supernovae. The idea that brightest stars in galaxies might all have the same luminosity and therefore be a good estimator of distance had been used by Hubble in his 1929 paper, but had fallen into disfavour. But in a series of papers published in 1979–80, the American astronomer Roberta Humphreys had shown that the brightest red supergiants in Local Group galaxies really did all seem to have the same luminosity. Sandage and Tammann used this method to try to establish the distance to two galaxies in the nearby Canis Venatici cluster in which Type I supernovae have occurred. These are stellar explosions where a white dwarf star in a binary system blows up because too much material falls onto it from its companion. Type I supernovae reach enormous luminosities, outshining for a few months the whole of the rest of the galaxy's starlight, and can be seen to great distances. Surprisingly, they always seem to reach about the same luminosity at maximum light and so make an excellent distance indicator. Sandage and Tammann's new Hubble constant was, once again, 50 km/s/Mpc. The luminosity of Type I supernovae at maximum light can be calculated from theoretical models of exploding white dwarfs and can also be estimated geometrically from direct observations of the expanding atmosphere of the supernova. Both these estimates agree reasonably well with the value derived by Sandage and Tammann. Type II supernovae, which are due to the explosion of massive stars (more than ten times the mass of the sun) at the end of their life, can also be used to estimate distances to galaxies, but they do not reach the same huge luminosities at maximum light, so can not be studied to such great distances.

Meanwhile a new team of astronomers had entered the distance scale fray. Marc Aaronson and Jeremy Mould of Steward Observatory began to use the Tully–Fisher method with the infrared luminosities of galaxies in place of the optical luminosities. They found that the correlation of a galaxy's luminosity at an infrared wavelength of, say, 1.6 microns, with its rotation velocity seemed to be even tighter than the relation at optical wavelengths. They were able to apply the method to clusters of galaxies at much larger distances than had been achieved with any other method except Type I supernovae. Barry Madore and colleagues had also shifted to the infrared in their studies of Cepheid variable stars. The correlation between the luminosity of the Cepheid and its period of pulsation showed less scatter at infrared wavelengths than in the optical band, and infrared observations reduced the problem of having to make corrections for extinction by interstellar dust. Their new distances to Local Group galaxies broadly agreed with those of Sandage and Tammann.

In 1982 I completed a first draft of my book on the distance scale and the Hubble constant, *The Cosmological Distance Ladder*, which I had begun thinking about at the 1976 Paris conference on 'The Redshift and the Expansion of the Universe'. Naturally, I sent the manuscript to Sandage, Tammann and de Vaucouleurs, among others. I met Gerard at a summer school at Varenna on Lake Como in the summer of 1982 where we were both lecturing. He had read the manuscript extremely closely and from his encyclopaedic knowledge of the

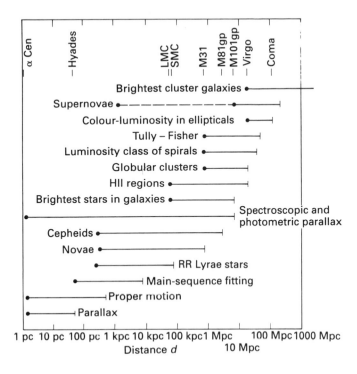

Fig. 4.6 The cosmological distance ladder, showing the range of distances over which different distance indicators have been applied. From The Cosmological Distance Ladder *(Rowan-Robinson, 1985).*

astronomical literature and of the history of astronomy produced hundreds of corrections and comments. Gustav wrote to me with a much shorter list of comments. Because my final figure for the Hubble constant was closer to his figure, he was basically happy with the book despite the fact that there were some criticisms of his and Sandage's approach. Allan Sandage is someone for whom I have always had the greatest admiration. His work has spanned so much of modern astronomy and he has always pushed the subject forward to its limit. I received no comments from him and when I spoke to him he claimed not to have read the book. But Gustav had already told me that 'Allan had been surprised to find the book was not as bad as he expected it to be'.

By 1982 a new generation of astronomers had taken up the challenge of the cosmological distance ladder and had developed a whole series of new distance methods. The conclusion of my book was that the future of the distance scale lay with the two new methods capable of reaching to substantial distances: the Type I supernova method and the Tully–Fisher method for spiral galaxies, both of which I have described above. Unfortunately, the two methods disagreed strongly with each other about the distances of galaxies. A new distance scale controversy had begun.

In 1987 Alan Dressler, Donald Lynden-Bell and their colleagues announced a new method which can also reach to substantial distances, the 'D-σ' method for elliptical galaxies. This depends on an empirical correlation between the diameter of galaxies and the average velocity of the stars in the galaxies. This method appears to give distances intermediate between those of the infrared Tully–Fisher and Type I supernova methods. Another development which favours an intermediate distance scale was the detection of novae, which are less violent eruptions of white dwarfs in binary systems, in galaxies in the Virgo cluster by Chris Pritchett and Sidney van den Bergh. However, the disagreements between the distance methods capable of reaching to large distances have still not been resolved today, so despite all the effort of the past twenty years we can still only conclude that the Hubble constant almost certainly lies in the range 40–100 km/s/Mpc, and probably lies in the range 50–90 km/s/Mpc, which corresponds to a maximum age for the universe of 11–20 billion years.

A certain strange star was suddenly seen

There have been three exciting developments in recent years which hold out great hope for the future of determining the Hubble constant accurately: the detection of a supernova in the Large Magellanic Cloud, the discovery of gravitational lenses, and finally the growing prospect of detecting individual bright stars in the galaxies of the Virgo Cluster.

'Behold, directly overhead, a certain strange star was suddenly seen . . . Amazed, and as if astonished and stupefied, I stood still.' With these words Tycho Brahe began his account of his discovery of the supernova, or exploding star, of 1572, the first in the history of humankind to have been the subject of serious scientific scrutiny. True, the Chinese astronomers had recorded the appearance of many supernovae from at least 185 AD onwards. But they seem merely to have noted their occurrence and drawn from them what astrological portents suited the times. Tycho observed his supernova night after night, recording its changes in brightness and colour, and demonstrating that it belonged to the allegedly unchangeable realm of the fixed stars. Incidentally, Tycho could hardly have missed the new star since it lay just above the W of Cassiopiea, one of the best known constellations in the sky.

Only 32 years later, in 1604, Tycho's disciple Johannes Kepler was able to study another supernova visible to the naked eye. But perversely no supernova visible to the naked eye was to occur for almost the next four centuries, although astronomers estimate that one occurs every fifty years or so in the Milky Way. The problem is that most of these are hidden from view by the dimming effect of interstellar dust. The first supernova to be discovered in an external galaxy, that of 1885 in the Andromeda galaxy, was found by telescope and just failed to make it to naked eye visibility. At that time the distance of the Andromeda system was not known and it was a further forty years before Hubble established the huge distance of this galaxy. Only then could the enormous luminosity of the 1885 event begin to become apparent. Fritz Zwicky coined the name 'supernova' for such events. Since then astronomers have regularly searched galaxies for supernova events and have been able to begin to estimate distances to galaxies using them as beacons.

The four-hundred-year dearth of bright supernovae explains the fever which gripped astronomers at the announcement by Ian Shelton on February 23rd, 1987 that he had discovered a supernova in the Large Magellanic Cloud. The LMC, as astronomers call it, is a dwarf irregular galaxy in close orbit around the Milky Way, at a distance of only 160 thousand light years. It lies near the south celestial pole and was first reported to Europe by Magellan's circumnavigating expedition of 1519–22. Thus it was the telescopes of the southern hemisphere which hastily and excitedly cancelled their planned observing programmes and turned to stare at the astronomical event of the century. The ultraviolet satellite IUE made key observations which helped to identify the star which had exploded. Theoreticians hastily ran new computer models for exploding massive stars which are responsible for supernovae like 1987A. Scientists in underground laboratories checked to see whether their high energy particle detectors might have seen the burst of neutrinos, massless and chargeless particles emitted when the collapsing core of the dying star crashes to form a neutron star. It is

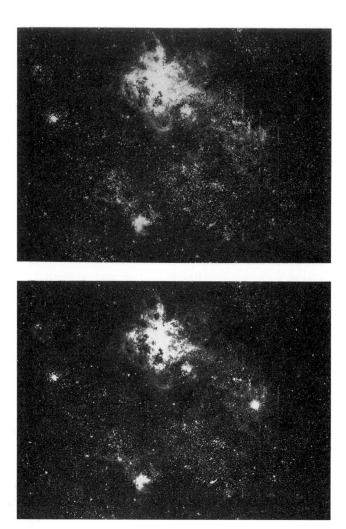

Fig. 4.7(a) Supernova 1987A in the Large Magellanic Cloud. The upper panel shows part of the Cloud before the supernova event, the lower panel shows the same region after the explosion (the bright star right of centre).

remarkable that the death of a star of twenty times the mass of the sun could be calculated so accurately. These models predict that most of the energy dissipated when the core collapses is released in a burst of neutrinos. These then interact with the surrounding surface layers of the star, driving a shock wave through the whole star. The temperature

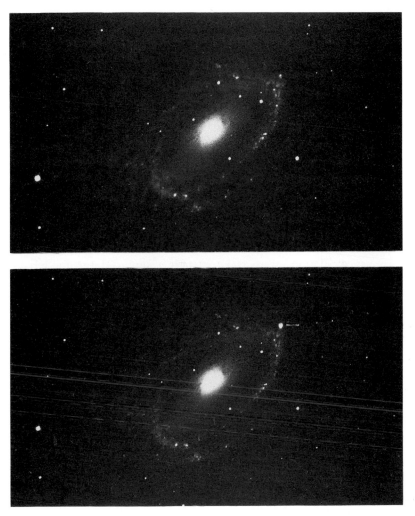

(b) A supernova in a distant galaxy.

in the shock wave is so great that a variety of rapid neutron capture reactions take place. This is where many of the elements and isotopes of the periodic table are manufactured, in particular where the radioactive isotopes are made. The most important reaction involves the formation of about one solar mass of nickel-56, which then radio-actively decays to cobalt-56, which in turn decays to iron-56. It is this process which releases most of the energy in the envelope of the star in the later stages of the development of the supernova event. Sure enough the neutrinos were detected in the amount expected and the light from the supernova, after reaching its rapid maximum, began its slow exponential decay, with a half-life of 77 days, due to the radioactive decay of cobalt-56 to iron-56.

Of course everything was not quite as simple as I have described above. When we received Ian Shelton's telegram, those of us with an interest in supernovae were soon estimating that it should become one of the brightest stars in the sky within a few days. Instead it reached only magnitude 4.5, one hundred times fainter than expected. The theoreticians were taken by surprise and had to go back to the drawing-board. It was discovered that the progenitor star, instead of being a huge red supergiant as expected for a dying massive star, was a much smaller blue giant that had lost most of its outer layers. Only when the supernova calculations were run for such a star, did the models begin to agree with what had been seen in SN 1987A. A key factor in the story is that the stars of the LMC have a much lower abundance of 'heavy' elements (elements from carbon onwards) than those in our Galaxy, because the LMC has been forming stars (where the heavy elements are manufactured) at a much slower rate than our galaxy. This lower heavy element abundance profoundly alters the evolutionary history of the stars in the LMC.

Once the right model for the supernova was found, it became a simple matter to measure its distance. Even the information given in Ian Shelton's February 23rd telegram on the temperature and expansion velocity of the supernova gas allowed a rough, back-of-envelope, distance to be estimated. This rough estimate agreed well with the distance to the LMC estimated by many other methods. More detailed observations and calculations by David Branch, David Arnett and others have confirmed this. Supernova 1987A therefore provides (after a few hiccups) a splendid validation of the supernova method of estimating distance and brings us closer to a definitive measure of the Hubble constant. Current estimates from supernovae lie in the range 50–65 km/s/Mpc.

As if one nearby supernova in our lifetimes was not enough excitement, on March 31st, 1993 an amateur astronomer in Madrid found a new supernova in the galaxy Messier 81, at 10 million light years the largest member of one of the nearest groups of galaxies to our own Local Group of galaxies. This discovery is bound to help pin down the distance scale and the Hubble constant, for M81 is a galaxy in which Cepheids have been studied, so the two distance methods – Cepheids and supernovae – can be tied together.

The discovery of gravitational lenses

The second of the new developments in measuring the Hubble constant involves a totally different piece of exotica of modern astronomy, the *gravitational lens*. Because, according to the General Theory of Relativity, gravity bends the path of light, a star or galaxy can act like a convex lens, bending the light around it to form an image of a source far behind the lensing mass. Gravitational lenses do not make as perfect images as glass ones however. The image of the source may be distorted to produce several images. The most common phenomenon is one or more circular arcs. If the alignment of source and lensing object is exact, then the image of the source becomes a ring. Many examples of such gravitationally lensed images, mostly quasars lensed by galaxies, have been discovered since the first was recognized in 1979.

The properties of gravitational lenses were first worked out in detail by the Norwegian astronomer Sjer Refsdahl in the 1960s. I

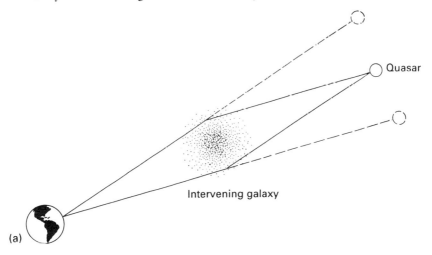

Quasar

Intervening galaxy

(a)

Fig. 4.8(a) Gravitational lensing of light from a distant quasar by an intervening galaxy. The light is bent slightly as it passes around the galaxy, generating two or more images of the quasar.

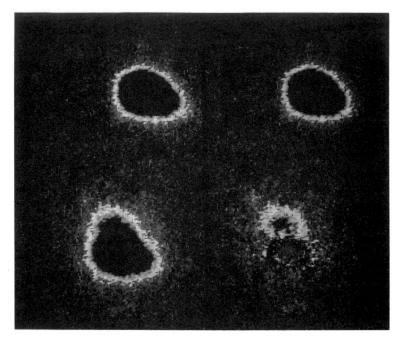

Fig. 4.8(b) The gravitational lens quasar 0957+561. The left-hand frame shows the two lensed images of the quasar, with the lensing galaxy merged into the lower image. In the right-hand frame the top image has ben subtracted from the lower image to show the lensing galaxy more clearly.

remember meeting him at the time and, like many other astronomers of the day, telling him this was far too fanciful an idea and that such lenses were unlikely to be found. But Sjer, a shy but determined man, persisted with his calculations and even showed how a gravitational lens could be used to measure the Hubble constant. All that was needed was for the background source to be variable and for these variations to be recognized in two different lens images, for which the light had taken different routes round the lensing galaxy. This phenomenon has now been seen and the first tentative estimates of the Hubble constant made. They lie in the range 50–75 km/s/Mpc and the accuracy is likely to keep on increasing as more examples are found.

The distance to the Virgo cluster

The final method, which is crucial for the final tying down of the cosmological distance ladder, involves identifying and measuring individual bright stars in galaxies in the Virgo cluster. The Virgo cluster is crucially important to the cosmological distance ladder because it is the nearest rich cluster of galaxies. Several thousand galaxies are concentrated into a volume not much larger than the three million light years occupied by the thirty galaxies of our Local Group of galaxies. Such clusters can be identified on photographic plates and imaging detectors to enormous distances. Although there is enormous disagreement between the supporters of different distance methods about the actual distance of Virgo, there is rather good agreement between the methods about the relative distance of clusters. If the Virgo cluster distance could be settled, most of the controversy about the distance scale and the Hubble constant would end.

Measurement of the brightest stars in Virgo cluster galaxies was to have been one of the great achievements of the Hubble Space Telescope, but the mirror aberrations mean that this goal will now only be possible when NASA's repair mission has been successful. There are claims that the goal has begun to be achieved with ground-based telescopes by groups working on Hawaii (led by Mike Pierce) and on La Palma in the Canaries (led by Tom Shanks). If they are right, the distance to the Virgo cluster is at the lower end of the current range of estimates and the Hubble constant might be as high as 90 km/s/Mpc. However, the brightest stars in galaxies are not a very reliable standard candle for distance estimates. We do not really know that the brightest stars in different galaxies always have the same luminosity. More compelling evidence would come if Cepheid variable stars could be detected in Virgo cluster galaxies, because Cepheids have a well determined relationship between pulsation period and luminosity, and so can be used to estimate distances accurately. To do this for Virgo galaxies is a harder task than studying the brightest stars because Cepheids are not among the brightest stars in a galaxy.

In a recent very interesting development announced in 1992, Allan Sandage and colleagues have used the Hubble Space Telescope to study Cepheids in the galaxy IC4182. Although three or four times nearer than the Virgo cluster, this galaxy is of special interest because in 1937 a Type I supernova exploded there. It is the nearest galaxy to us where this has happened in (relatively) modern times.

Sandage and his colleagues studied some 27 Cepheids in IC4182 and were able to measure an accurate distance to the galaxy of 15 million light years. This gives a direct estimate of the luminosity of the 1937 supernova, which they can then use to estimate the distances to other much more distant galaxies in which Type I supernovae have occurred. From the recession velocities of these galaxies they can then determine the Hubble constant, which they find to be 51 km/s/Mpc. This is the most convincing evidence to date that the Hubble constant may lie at the lower end of the currently accepted range.

As we shall see later, other developments in cosmology strongly point to a Hubble constant at the lower end of the allowed range and I shall be surprised if the ultimate value lies outside the range 50–70 km/s/Mpc. This would give an expansion time-scale for the universe of 14–20 billion years. Because gravity has the effect of slowing down the expansion, the age of the universe is always less than the expansion time-scale today. In the simplest model of the universe, the age would be two-thirds of the expansion time, 9–13 billion years. Currently, the best estimate of the oldest stars in our Galaxy is 13 billion years, with an uncertainty of about a billion years either way, and radioactive dating gives ages in the range 10–15 billion years. Everything could just fit together nicely. But there is now a really exciting period ahead when the accuracy of these different measurements may reach a level at which we really begin to learn something new about the universe.

The Big Bang universe

We now turn to explore in more detail what the Hubble law tells us about the universe we live in. Edwin Hubble's discovery that the galaxies are receding from us at a rate that increases with distance showed that we live in an expanding universe. Extrapolating the expansion backwards in time points to an origin in a single event ten to fifteen billion years ago. However, the origin of the universe in an explosive 'Big Bang' was not the only possible explanation of the Hubble law. Herman Bondi, Tommy Gold, and Fred Hoyle had in 1948 put forward the idea that the expanding universe is in a steady state, with new matter being created continuously to maintain the average density at a constant level.

It was the discovery of the microwave background radiation in 1965, which I describe in detail in Chapter 8, that showed that the universe really did have its origin in a Big Bang and that the early phases of the universe were dominated by radiation. The universe was, in fact, born in a blaze of light. There were three pieces of evidence that convinced most astronomers of this and gradually destroyed any possible alternative models. First, the impressive smoothness and isotropy of the background radiation did not fit in with any origin in the solar system or the far from smooth distribution of galaxies. Secondly the shape of its spectrum, the way that the intensity changed with frequency, perfectly matched the Planck 'blackbody' spectrum expected from a phase of matter and radiation in perfect thermal balance with each other. Material which is a perfectly efficient absorber or emitter of radiation is described as black. A terrestrial realization of a blackbody is the interior of a cavity maintained at a very uniform temperature like a furnace. The surfaces of the stars are, approximately, blackbodies. The German physicist and founder of quantum theory, Max Planck, showed that the spectrum of radiation emitted by a blackbody has a shape, peak wavelength and intensity which depends only on the temperature of the body. When we see such a spectrum we know that we are seeing a phase where are in thermal balance with each other. Since matter and radiation are certainly not in thermal balance with each other in the universe today, the blackbody spectrum of the microwave background radiation shows that there must have been such a phase in the past and confirms this radiation as the relic of the fireball phase of a hot Big Bang universe. Finally, the clinching piece of evidence for the hot Big Bang picture is that the observed abundances of the light elements, helium, deuterium and lithium, which can not be explained by stellar processes, agree well with the predictions of the model.

The microwave background radiation is remarkable for its uniformity around the sky, except that it looks slightly brighter in one direction and slightly dimmer in the opposite direction, by about one part in a thousand. The simplest intepretation of this slight non-uniformity is that our Galaxy (along with other nearby galaxies) is moving through space at a speed of about 600 km/s. Recently, a group of colleagues and I have used 2400 galaxies found by the IRAS satellite to map out the distribution of mass in the universe within 300 million light years. We found that the motion of our Galaxy can be accounted for simply as the result of the gravitational pull of a dozen or so large galaxy clusters within the volume we have surveyed. I will describe this discovery in more detail in Chapter 9.

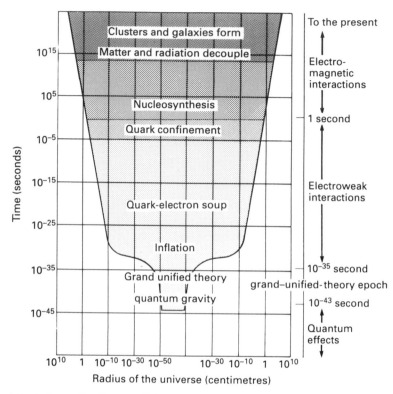

*Fig. 4.9 Schematic picture of the evolution of the universe. In the 10⁻⁴³
seconds following the Big Bang, quantum effects dominate and the four
fundamental forces (electromagnetism, weak and strong nuclear forces and
gravity) are believed to have been unified in a single force. First gravity
separates out, leaving the other three forces as a 'Grand Unified Force'.
When the strong nuclear force separates from the 'electroweak' force 10⁻³⁵
seconds after the Big Bang, inflation begins. The matter in the universe
consists of a 'soup' of quarks, which are the building blocks of protons and
neutrons, and electrons, but the dominant form of energy is radiation. When
the universe is one second old, the quarks bind together to make protons and
neutrons, and the weak nuclear and electromagnetic forces separate.
Nucleosynthesis begins and continues until the universe is about three
minutes old. When the universe is 300,000 years old, the matter cools
sufficiently to become transparent to radiation and galaxies and clusters of
galaxies can begin to form.*

Once the motion of our Galaxy through space is allowed for, the microwave background radiation looks the same in every direction to within one part in a hundred thousand, a remarkable uniformity. Yet the microwave light we are seeing was emitted by matter so far away that the particles of matter in the different directions could not have had any influence on each other by any physical mechanism limited by the speed of light. The microwave background light we detect today was last emitted by matter a few hundred thousand years after the Big Bang. At that time the matter could have been influenced by other matter within a few hundred thousand light years. To see what volume of the universe that corresponds to today, we have to allow for the fact that the universe has expanded by a factor of a thousand since then. Today the volume within the 'horizon' of the emitting matter would correspond to a region of diameter a few hundred million light years, quite a small volume compared with the twenty billion light years that separate two regions on opposite sides of the sky. How did all these isolated pieces of matter 'know' they had to be the same? This is known as the *horizon problem* and is a real paradox, which in the simple Big Bang model can only be resolved by saying that the universe was born with this uniformity.

This brings us to a second problem with the Big Bang model, one that has been at the heart of cosmological research for the past decade and is the main point of this book. The galaxies, the stars and we ourselves demonstrate that the universe today is very far from being uniform. How did these fascinating structures form from such a smooth and uniform initial state? The presumption has generally been that gravity plays the key role in building up the amplitude of small irregularities in density. We shall see that there is a problem in forming the galaxies and clusters of galaxies in the time available. Gravity acts too slowly in amplifying the density irregularities. Then we have to ask, how did these density irregularities appear in the first place? In the simplest General Relativistic cosmological models we again have to resort to assuming that small non-uniformities were present at the birth of the universe, and were of just the right form to evolve into the structures we see today.

Cosmic inflation

Both of these problems are claimed to be solved in the *inflationary*

Fig. 4.10 Alan Guth of the Massachusetts Institute of Technology (MIT), inventor of the inflationary model of the universe.

model, first put forward by Alan Guth (now at the Massachusetts Institute of Technology) in 1981. I will give a more detailed account of this theory in Chapter 10, but for the moment here is a brief outline. In the inflationary picture, the whole universe that we see today has inflated from a single infinitesimal 'seed'. The distant pieces of matter that we sample with the microwave background radiation were all close enough to influence each other before the inflation began. Where does the energy for the colossal expansion which is supposed to take place during the brief inflationary period come from? Mathematically, the universe behaves as if the universe suddenly acquires a huge value for Einstein's 'cosmological repulsion' force. Einstein introduced this force to allow him to make a static model for the universe, with the cosmological repulsion balancing the self-gravity of the universe. Later he and other relativists took the view that this term in the equations should be zero. Guth revived the idea of the cosmological repulsion and gave it a new interpretation. It corresponded to the energy-density of the vacuum. According to particle physics theories, the vacuum may, in the extreme conditions of the very early stages of the Big Bang, find itself in an anomalous state of huge energy density. Paradoxically then, the energy for the exponential expansion is postulated to come from the vacuum itself. This seems almost like a return to Aristotle, who argued that fear of the vacuum drove a projectile through the air.

The trigger for the inflation, in many versions of the theory, is a phase transition of the matter in the universe, associated with what is called the Grand Unified epoch. This is the moment when the two basic forces of physics, the strong nuclear force (which holds the nuclei of atoms together) and the 'electroweak' force (a unification of the electromagnetic force and the weak nuclear force responsible for radio-activity), separate. Before this event, which is supposed to occur an incomprehensible 10^{-35} seconds after the instant of the Big Bang, these forces comprise a single Grand Unified Force. The detailed physics of the phase transition can also explain the origin of the small irregularities in density which we believe must be present in the early universe for galaxies and clusters of galaxies to grow from. As a further bonus it can explain why we live in a universe dominated by matter rather than anti-matter.

One requirement of the inflationary model is that the universe contain a very high proportion, perhaps 97%, of matter in some invisible, exotic form, for in this model the universe would have the 'critical' density which distinguishes an open universe, which will keep on expanding for ever, from a closed universe, which will eventually recol-lapse to a 'Big Crunch'. Visible matter in galaxies accounts for only 1% of the critical density. Including the dark halos believed to surround most galaxies, perhaps 10% of the critical density is accounted for. The inflation scenario predicts an even more all-pervasive distribution of dark matter. Interestingly the IRAS galaxy survey I mentioned above does point to such a pervasive distribution of dark matter (see Chapter 9). But what this dark matter is remains a mystery. And even if we could detect the dark matter it would not really prove the correctness of the inflationary scenario. In fact I find it quite hard to see how this scenario can be tested observationally, though theoreticians like Stephen Hawking claim the microwave background 'ripples' are direct evidence for inflation. At the moment the model of the inflationary universe, despite the immense effort being put into it by the theoreticians, remains a fascinating but rather metaphysical speculation.

It is interesting to compare how well inflation theory performs, compared with the General Theory of Relativity, against the criteria I gave in the previous chapter for a good new model: consistency with successful parts of past theories, explanation of previously unexplained phenomena and prediction of entirely new phenomena. Inflation is consistent with the pre-existing model of the universe, the standard Hot Big Bang model, because when inflation ends the universe continues as it did in the standard Big Bang model. It does explain phenomena not understood in the old model because it solves the horizon problem and explains how the very small density fluctuation, which would eventually result in galaxies and clusters of galaxies, could have arisen in the very early universe. The problem, though, is that there are no predictions which are uniquely associated with the inflationary model. In most versions of the model (but not all) the density of the universe today would be very close to the critical value, which distinguishes a universe which will expand for ever from one in which the expansion will halt and be reversed. However, the critical value is also a very natural value to assume in the old standard Big Bang model and several theoreticians had proposed this long before the invention of inflation theory. Some versions of inflation, but not all, predict a particularly simple form for the distribution of the strength of the initial density fluctuations on different scales. But such a simple distribution was in fact suggested in the standard Big Bang model 15 years before the invention of inflation. Unless inflation theory can come up with some more concrete predictions, it will remain an interesting idea, but it is not yet a good scientific theory.

The years of building IRAS

I AM OLD enough to remember when the conquest of space was still a fantasy. In the radio days of my childhood, I remember being glued to the radio each week for 'Journey into Space'. Later, in my teens, there were the extraordinary achievements of the Soviet space programme, the first Sputnik, the orbiting of the dog Laika, and finally Yuri Gagarin himself, the first human in space. I was moved watching the BBC documentary 'Red Star in Orbit' which showed the Soviet astronauts' tradition of visiting Gagarin's office for a few moments of quiet reflection before leaving on a mission. His office with the clock stopped at the moment of his death.

Another vivid memory is the meeting of the Royal Astronomical Society in London one Friday in 1966 at which a slide was shown of the first picture of the moon's surface from NASA's Surveyor I camera. Suddenly there was a real landscape with rocks, dust and distant hills. The space exploration of the solar system during the past few decades has been a liberating experience. But to me, the truly great voyages of discovery, the ones that compare in importance to human history and culture to those voyages of the navigators in the fifteenth and sixteenth centuries, are those that we have made in the past thirty years with our telescopes, opening up the invisible wavelengths. Space missions have played a crucial part in this era of discovery.

The growth of infrared astronomy

The development of solid-state infrared detectors in the 1960s allowed infrared astronomy to grow rapidly in the atmospheric windows available to ground-based astronomers. These bands are at wavelengths of 1.2, 1.6, 2.2, 3.5, 5, 10 and 20 microns (1 micron, or micrometre, is one millionth of a metre, or one thousandth of a millimetre). Because water vapour absorbs and radiates strongly at the longer infrared wavelengths (3 microns or longer), the longer wavelength windows are usable only on good, dry mountain sites. California and Arizona (and later Hawaii) became the main centres for infrared astronomy.

In 1965, Gerry Neugebauer and Robert Leighton of California Institute of Technology embarked on a three-year survey of the sky at a wavelength of 2.2 microns with a small 1.6-metre telescope on Mount Wilson, near Los Angeles. The data, which consisted of miles of paper with the wiggly signals from the detectors plotted on them, were inspected visually by a team of part-time assistants, many of them students. Gerry Neugebauer decided to take a very cautious line about what would be published and only very clear, strong signals were included in the catalogue. Most of the sources found in the survey were stars, but many of them were unexpectedly bright in the infrared because they are surrounded by clouds of dust which absorb the visible light from the star and re-emit it at infrared wavelengths.

In 1967, Gerry Neugebauer and his research student Eric Becklin discovered a new kind of infrared source at a wavelength of 10 microns in the constellation of Orion. The Becklin–Neugebauer object turned out to be a very young star, which had only just formed at the centre of a thick cloud of dust and gas, close to the four Trapezium stars which make up the central 'star' of the sword of Orion. The characteristic temperature of the source was only 600°K (degrees Kelvin, or absolute), far cooler than the 2000°K surface temperature of the coolest stars known. (Degrees Kelvin are the same as degrees Centigrade, except that they have a different zero point. 0°C corresponds to 273°K.) Soon afterwards Frank Low and Douglas Kleinmann, of the University of Arizona, found an even younger and cooler object in the same part of the sky, a cluster of stars in the process of formation, with a characteristic temperature of only 60°K. It was becoming clear that moving to longer wavelengths would yield new kinds of astronomical objects and insights into new astrophysical processes.

The tendency to be studying a cooler universe as we move to the longer wavelengths, and lower frequencies, of the infrared follows from the fact that the energy carried by a photon, or particle of light, is directly proportional to the frequency of the light. Radiation from a hot gas cloud like a star, or from a thick cloud of dust, tends to be emitted by thermal processes in which the photons acquire an energy comparable with the average energy of the atoms or molecules of the gas or dust. And the average energy of the atoms or molecules is exactly how we define the temperature of gas or dust. Thus the lower frequency photons of the infrared carry less energy than optical photons, and tend to have been emitted by gas or dust atoms or molecules with lower average energy, and hence at lower temperature. If the thermal balance between the matter and radiation in the emitting region were exact, then the radiation would have a blackbody spectrum (see p.63), and the peak intensity of the radiation would be precisely related to the temperature of the matter. Many of the astronomical objects we see in the infrared are only approximately blackbodies, but the relationship between the wavelength we are observing and the temperature of the matter remains reasonably accurate. Thus as we move towards longer infrared wavelengths we tend to see an ever cooler universe.

In the late 1960s, Frank Low developed detectors capable of detecting much longer wavelength, far infrared radiation, at wavelengths 100 microns to 1 mm. These allowed him to look for even cooler sources, at only a few tens of degrees above absolute zero (the absolute zero of temperature, zero degrees Kelvin, is where motions of atoms cease altogether, and corresponds to −273 degrees Centigrade). These wavelengths are important for measuring and understanding the total energy output of galaxies and the role of interstellar dust in them. A significant fraction of the visible light emitted by stars in spiral galaxies like our own is absorbed by interstellar dust and re-emitted at far infrared wavelengths. Low used these detectors first with ground-based telescopes in the wavelength window around 1 mm, through which some radiation penetrates the earth's atmosphere to reach the ground. He then switched to observations with a small telescope on a converted executive jet, in which he could fly above most of the earth's atmosphere, at wavelengths of 100–350 microns. However progress was slow and very few objects could be detected at these wavelengths.

Between 1971 and 1974, Steve Price and Russ Walker, working at the US Air Force Geophysics Laboratory, carried out a rocket survey of the sky at wavelengths of 4, 11, 20 and 27 microns. Unfortunately they did not take the careful approach of the 2-Micron survey and published a preliminary catalogue which contained many spurious sources. My research assistant Stella Harris and I found that the genuine sources could almost all be characterized as normal stars, stars surrounded by a cloud of dust (or circumstellar dust shells) or star-forming clouds. We set out to make detailed models for the infrared spectra of all the reliably detected circumstellar dust shells in this survey, using computer codes I had developed for analysing the flow of radiation through a dust cloud.

During the 1970s many different groups worked on far infrared astronomy from telescopes carried by balloons, rockets or aircraft, for example the groups led by William Hoffmann at the University of Arizona, by Jim Houck and Martin Harwit at Cornell, by Dick Jennings at University College London, by Ed Erickson at NASA-Ames, California, and by Al Harper at the University of Chicago. By the end of the decade it was clear that many galaxies emitted much of their radiation at far infrared wavelengths.

First plans for IRAS, the Infrared Astronomical Satellite

In 1973 plans began to be put together for a new astronomy space mission. The idea, originating in the Netherlands, was for a survey mission at far infrared wavelengths, between 10 and 100 microns. As I mentioned above, these wavelengths are crucial for understanding the energy output of galaxies and for studying interstellar dust in them. Much of the light radiated by stars, over 99% in some galaxies as we later found, is absorbed by interstellar dust. The dust grains tend to attain temperatures of about 20–50°K and re-radiate the energy they have absorbed at far infrared wavelengths. The typical size of these grains is 0.01–0.1 microns and they are found in two varieties, silicates and amorphous carbon (like very fine sand and soot, respectively).

The problem is that the far infrared wavelengths where the dust grains radiate are inaccessible from the ground because of the strong absorption of the earth's atmosphere, mainly due to water vapour and carbon dioxide. Hence the need for a space mission, to get a telescope with sensitive far infrared detectors above the earth's atmosphere. For both financial and technical reasons the Dutch needed additional partners for this mission and they approached NASA. It turned out that American infrared astronomers led by Gerry Neugebauer and Frank Low had already proposed an infrared survey mission to NASA as part of the Explorer programme. This was a programme of small satellites, with a tight budgetary limit, which were supposed to be launched at the rate of one or two per year. The combined Dutch-US proposal began to look feasible and a team to define the mission was formed in 1975. The Dutch were also keen to have the UK as minor partners in the mission, so that the European contribution could be 50%. They approached the UK Science Research Council, who despatched Dick Jennings, Peter Clegg and Phil Marsden, three leading UK far infrared experimentalists, to a key meeting to define the mission in Washington. The mission was to become one of the most successful space astronomy missions of all time, the Infrared Astronomical Satellite, or IRAS.

In order to secure funding for UK participation, a scientific case for the mission had to be proposed to the Science Research Council. Dick Jennings, Peter Clegg and Phil Marsden called a meeting at Sussex University in Brighton of interested people in the astronomical community and out of this the case was written. The meeting divided up into a number of parallel sessions on different areas of astronomy. I chaired the session on extragalactic astronomy and cosmology, and wrote that part of the case. Although it was clear that the mission would be important for studying individual galaxies, the idea of using this survey for studying the universe as a whole seemed rather speculative, since very little progress had been made in cosmology at infrared wavelengths. But although IRAS made many exciting discoveries, it turned out that it was for cosmology that IRAS became most famous. In due course the Science and Engineering Research Council (SERC), as it became, agreed to fund a 10% UK involvement in IRAS and signed a Memorandum of Understanding with NASA and the Dutch funding agency, (NIVR).

It takes many years to design, build, test and launch a spacecraft. Astronomical missions are especially complex and demanding, requiring very stable platforms, accurate pointing and control, and sensitive detectors. IRAS was to be the first launch into space of a telescope

*Fig. 5.1 The IRAS
satellite
(a) artist's impression in
orbit,
(b) in the laboratory
before launch,
(c) exploded view of
satellite,
(d) the array of detectors
in the focal plane.*

(b)

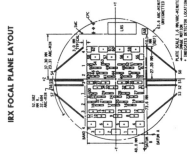

IRX FOCAL PLANE LAYOUT

(d)

(a)

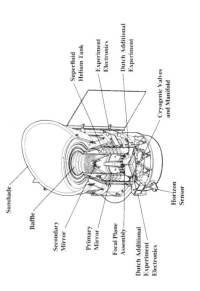

Sunshade

Baffle

Secondary
Mirror

Primary
Mirror

Focal Plane
Assembly

Dutch Additional
Experiment
Electronics

Superfluid
Helium Tank

Experiment
Electronics

Dutch Additional
Experiment

Cryogenic Valves
and Manifold

Horizon
Sensor

(c)

cooled to very low temperatures. The point of the very low temperature was to eliminate the infrared emission from the telescope and reduce the thermal noise from electrons in the detectors to as low a level as possible. IRAS essentially consisted of a telescope with a 60cm diameter mirror, sitting inside a vast thermos flask or dewar. The coolant in the dewar was to be superfluid helium, cooled to $-270°C$, only 3°C above the absolute zero of temperature. The helium would slowly evaporate past the detectors and telescope structure, maintaining them at a very low temperature. Eventually, though, the helium would all have evaporated away and the useful astronomical life of the mission would have ended. When a gas is cooled and becomes liquid, and then continues to be cooled at a low pressure so that it cannot solidify, it can become a 'superfluid', almost without viscosity. Before IRAS, superfluids had never been used in space. Thus in several respects, IRAS was a daring step forward in space technology. The common gossip at that time was that the military had made several attempts to launch cryogenically cooled telescopes (designed to search for missile launches, not galaxies, of course) but without success.

Many hundreds of people are involved in a space mission, all with very different skills. At different phases of the mission, different types of people step forward to play the key role. All have to collaborate to make sure that their contributions mesh together for the common goal. One of the first steps is to decide what the scientific goals of the mission are. In this case the main goal was to make a survey of the whole sky in the far infrared waveband, at four separate wavelengths between 10 and 100 microns. In this waveband we are especially sensitive to material with temperatures in the range 30–300°K ($-240°C$ to 30°C), so we are looking at the cool universe of radiation from interstellar dust grains, rather than the universe of stars that we see in the visible waveband. The human body radiates at about 10 microns, so when you feel the radiated warmth of other human bodies in a crowded room or train, your body is behaving as a crude 10 micron detector.

Because of the enormous cost of space missions, it is common today for space astronomy missions to be international projects shared between several nations. The subdivision of tasks for IRAS was that the Dutch were to build the spacecraft, the British were to provide the ground station and 'quick-look' data analysis facility, while the Americans took on the challenging tasks of building the dewar and the telescope, and also the provision of the final data analysis facility. NASA also provided the launch. The British and Dutch were also free to build extra instruments to go at the focal plane of the telescope, in addition to the four survey channels, chosen to be 12, 25, 60 and 100 microns in wavelength. The Dutch provided a photometer for making sharper images at 100 microns and a low-resolution spectrometer for the wavelength range 8–20 microns. The British set out to provide an additonal wavelength channel at 200 microns wavelength, but had not solved the technical problems at the moment the design had to be fixed, and so abandoned this idea. Ironically, it turned out that the delays associated with the manufacture of the telescope meant that there would have been plenty of time to solve these problems.

The years of the windowless rooms

A major feature of the years 1976 to 1983, when IRAS was finally launched, was the science-team meetings twice a year at which the progress of the mission would be reviewed. These meetings alternated between Europe and the US. The US meetings were held at one of the three NASA centres involved in the project: Goddard Space Flight Center in Maryland, where the original planning for the mission was carried out, Ames Research Center in Moffett Field, California, which was responsible for the design and construction of the telescope, and Jet Propulsion Lab (JPL) in Pasadena, California, where the final assembly and testing took place, and where the data analysis facility was to be located. I came to know the Saga Motel, Pasadena, and the windowless meeting rooms of JPL all too well. The meetings in the Netherlands were sometimes at the Leiden Observatory and sometimes at the University of Groningen, the UK meetings generally at Appleton Laboratory and later when that merged with the Rutherford Laboratory, at the combined Rutherford–Appleton Laboratory (RAL). The science team meetings were chaired jointly by the US team leader, Gerry Neugebauer, and the Dutch team leader, Reinder van Duinen. Despite some tensions they did an excellent job of holding the team together through some difficult times. Shortly before launch, Reinder moved off to a senior job with the Dutch aerospace company Fokker, and was replaced by Harm Habing of Leiden University.

Gerry Neugebauer and Frank Low were the two dominant characters at IRAS science team meetings. Gerry was small in build, often anxious and on edge, and he worried about every single detail of the mission. If there was a problem, he would drive someone or other mad until they fixed it. Of all the scientists in the IRAS team, Gerry was the one who most shared my desire to create a really reliable and complete survey of the far infrared sky. If you reached his inner circle of fixers, on Sundays you got invited to lunch at a seedy Mexican restaurant in Altadena that he loved. The success of IRAS owes a great deal to Gerry and his determination to get things right.

Fig. 5.2 (a) Gerry Neugebauer,   *(b) Frank Low.*

Frank Low was different in build and personality. He was a large man in every way and had sublime confidence in his own intuition and experience. He was often proved right. He did not think much of the IRAS Point Source Catalog, to which I was devoting most of my time (that was one judgement of his that I think he will agree turned out to be mistaken), and was mainly interested in the IRAS maps of extended infrared emission. He and Gerry both had explosive tempers, held on rather short fuses, so meetings could get rather fiery. Neither of them had much respect for, or interest in, theoretical arguments. In the end, I came to respect and like them both, and they were both, in their different ways, kind and generous to me.

There was a considerable difference between the US and Dutch-British science teams. The US team, which had been selected by competition in the US community, comprised almost entirely infrared experimentalists, people whose prime love and expertise was building instruments and making observations. The Dutch and British teams, consisting of a mixture of scientific agency appointments and enthusiastic hangers-on (like myself), had a much higher proportion of theoretically minded astronomers. Thus while the Americans dominated the design and construction phases of IRAS, the Dutch and British came to play an increasing role in the data analysis and scientific analysis of the mission.

Although I was not formally a member of the IRAS science team until just before the launch of IRAS, I was assured that this would be sorted out later. Since I had no guarantee that I would see and be able to use the IRAS data until this situation was formalized weeks before launch, I was taking a risk that these years of effort would not lead to anything. The mission seemed so important that I thought it was worth the risk, though it was a tense time for me. One of the problems was that the US team did not want to enlarge their own science team, so they saw no reason why they should agree to the Europeans enlarging theirs. In the end all hands were needed at the pumps.

The group at Appleton Laboratory, as it then was, worked on the design of the ground station, the major UK commitment to the project. One of the first tasks of the UK team was to design the quick-look data analysis system. The purpose of this was to check out the survey data each day and see whether any of it needed to be redone. For this we needed to have an idea of how many sources we would see with IRAS and this was something which I happily accepted as a task. I took what was known about different types of astronomical object – stars, dust clouds and star-forming regions in the Milky Way, and galaxies – at near infared wavelengths, 2–10 microns, and extrapolated to the longer far infrared wavelengths to try to come up with some rough estimates. A few sources had already been observed at far infrared wavelengths from high-altitude aircraft or stratospheric balloons. In a report I circulated to the IRAS team in 1978, I concluded that about 100,000 stars would be seen, predominantly those surrounded by shells of dust, and that about 20,000 galaxies would be found, mostly 'emission-line' galaxies in which strong star formation was going on (these have become known as starburst galaxies), but with a 5–10% sprinkling of Seyfert galaxies. Seyfert galaxies are galaxies with an active nucleus which appears as a bright starlike core on a photographic plate and in which very rapid motions, up to tens of thousands of kilometres per second, are seen. They are closely related to quasars (see Chapter 11) and are believed to be powered by a massive black hole in the centre of the galaxy. My prediction of the total number of galaxies was heavily based on the 10 micron studies of George Rieke and Frank Low, of the University of Arizona. These predictions, though crude, turned out to be surprisingly close to the truth.

Another phenomenon that I discussed in my 1978 report was the problem of filamentary interstellar dust clouds which would emit strongly at 100 microns. Allan Sandage had noticed such filaments of dust at high galactic latitudes a few years before and suggested that the dust in our Galaxy has a 'cirrus-like' distribution. I estimated how much far infrared emission we might expect to see from this dust. Although my memo stated the possible problem of the infrared 'cirrus' at 100 microns fairly clearly, we all rather forgot about this issue (myself included) and were quite surprised when it turned up in the survey five years later.

Pre-launch problems

Things did not go smoothly with the construction of IRAS and several times the mission was close to cancellation. As the cost of the telescope and dewar construction steadily mounted, it became clear that NASA's financial contribution to the mission was first doubling then almost tripling (from $US37 million to $US92 million). An expensive test and calibration facility built at NASA-Ames never worked, because of the problem of establishing a good enough vacuum. The prime contractors for the detectors, Rockwell, managed to destroy about a third of the detector preamplifiers accidentally during test. It was decided to change the preamplifier design to one which had been advocated by Frank Low for some time. JPL were given the task of rebuilding the detector array. A 'tiger team' was appointed to carry out this task and managed to complete it by the end of 1981. As launch approached, however, a whole panel of detectors at 25 microns was failing to work because of a short-circuit to earth. Jim Houck, a US team member from Cornell, had the brilliant idea of switching the polarity of the detectors and making a new earth contact, which saved the panel. The dewar appeared to have a cold leak and not to be reaching the correct temperature. There was a fear that the cryogen would boil away in a matter of days instead of the planned one-year lifetime. The dewar cover, which had to sit on the front of the dewar during launch to protect the telescope and detectors and keep them cold, failed to separate from the dewar during tests. For about three years the launch date remained steadily two years away, always slipping further on. It was difficult for the Dutch and British, who had completed their tasks on time, to keep their teams of experts together during the delay. Eventually, the telescope was shipped to Holland for integration into the spacecraft and the complete IRAS satellite was then shipped back to JPL, Pasadena, for final testing.

In November 1982 the science team met and heard a review of the current status of the satellite. Things did not sound good. The detectors were much noisier in test than expected. Several detectors appeared to be 'dead' and did not work at all, which would leave holes in the survey. The estimated lifetime of the mission had come down from the planned 460 days to only 221 days, barely enough to carry out the survey. My estimate of the number of sources that would be detected had now come down from hundreds of thousands to a thousand or so, mostly in our galaxy. The science team voted on whether to recommend that the launch of IRAS go ahead, or whether the telescope should be opened up and the bad detectors replaced (this would have involved a delay of months). The vote was exactly evenly split.

At this point the NASA managers made one of their famous snap decisions, in which the recommendations of scientists and engineers are overruled in favour of management intuition. It seems to have been one of these snap management decisions that resulted in the launch of the Shuttle Challenger and the subsequent disaster, despite warnings that the rockets had been subjected to excessively cold temperatures in the days before launch. In the case of IRAS, NASA's intuition worked wonderfully well. NASA decided to launch IRAS, ready or not. The launch date was set for January 25th 1983.

I had already decided to spend the mission based at JPL in order to work on the final analysis of the data and the preparation of the survey catalogues. I was the scientist in charge of the detection subsystem, the package of software which examined the streams of data from the detectors and extracted the real astronomical sources of radiation. I worked closely with Tom Hibbard, the software engineer at JPL who had written the computer programmes. Another key person was John Fowler, the engineer who had written the software which took the many different detections of each source and combined them together into a best estimate of the brightness and position of the source. John was an excellent mathematician and his software always had a touch of style. He was also a strong character and could not be intimidated by the NASA management. At one time the US military began to take an interest in IRAS because its telescope would get sightings of all the satellites in higher orbits than IRAS, including some which they perhaps did not know about. John was asked to write some software which would recognize satellite crossings in the data, so that this information could be passed on to the military. He simply refused point blank to do so. I believe this task was eventually carried out by other less robust souls some years after the mission, as part of the Star Wars programme.

Just before the launch of IRAS I set out to Pasadena for a four-month stay. The idea was that this would be sufficient time to sort out all the problems with the software and set the catalogue production firmly on its way. My family was to join me in the middle two months. As I set out for Pasadena, others were travelling in the opposite direction to work at the ground station at RAL. I rented the house of Conway Snyder, a space old-timer whose speciality was writing computer routines for commanding the satellite into specific tasks like making a map of an area of the sky. While he worked at RAL in Chilton, Oxfordshire, my family and I lived in his large wooden bungalow in the Los Angeles suburb of La Crescenta.

The launch of IRAS

The launch itself was an unforgettable experience. We were driven out from Pasadena in coaches to Vandenberg Air Force Base. Of course this is not what you or I have in mind when we think of an 'air force base'. It is in fact a missile base and is the home of the Minuteman intercontinental missile. Over the same weekend as the IRAS launch, the American peace movement was organizing a protest at the Vandenberg Base. As a long-time opponent of nuclear weapons myself, I could not quite imagine how I would cope with meeting a picket by anti-nuclear protestors. In fact we were taken in by a back way and never saw any protest. For some of us there is a serious moral dilemma about working with NASA, because they are also heavily involved in military projects. One of the great features of the European Space Agency, which I hope will never be tampered with, is that its charter forbids its involvement with any kind of military work.

We were driven to a patch of open ground about one mile from the launch-pad. The huge white Delta rocket could be clearly seen in the distance, with its precious cargo perched on the top, painted with the brightly coloured flags of the three nations. The scientists with key responsibility for the detectors and telescope were in the control room in the distance monitoring the health of their babies via electronic read-outs. To my surprise Gerry Neugebauer joined the rest of us not directly involved in the hardware, out in the open. Perhaps he could not bear to follow the details of the launch too closely. A tense man at the best of times, he must have been tied in knots. It is a nerve-wracking moment and several scientific satellites have failed at launch, either by not reaching the right orbit, or by not switching on once there. A loudspeaker on a pole gave us a typical breezy NASA style of commentary on the launch.

> *It is launch minus 30 minutes and counting. The range controller has received confirmation from the coastguard that the launch path is free of shipping.*

> *It is launch minus 5 minutes and holding. We will be holding for one hour while all systems connected with the launch are checked through. This is normal practice for a Delta launch. You might be interested to know that the Delta team have made over 200 successful launches, with a 97% success rate.*

> *'The launch officer reports that all systems are ready. We will now resume the count-down. Launch minus 5 minutes and counting.*

> *Launch minus 30 seconds and counting.*

It is hard to convey the stillness in the dusk before the launch of a satellite that a group of you have spent seven years of your life preparing. At the moment that night falls, as the terminator that parts day from night crosses us, the sky lights up in an extraordinary flash, followed seconds later by a steady roar. The brilliant fireball spreads out along the ground. Still the rocket does not move. Then slowly it lifts and separates itself from the vast cushion of smoke. The brilliant flash of the first stage chases the rocket upwards. There is a second flash, less brilliant. Has it blown up? I hear Gerry Neugebauer say 'Christ' through clenched teeth and walk away. In fact the second stage has ignited perfectly. A few minutes later it has gone. It is a stunning and moving experience. I know this is a historic moment in my own life at least.

Fig. 5.3 Launch of IRAS, January 23rd, 1983.

Later there is a party in a huge sprawling house in the Air Force hamlet of Lompoc. Scientists, software engineers, industrial contractors, managers, the space programme professionals who integrate space missions together, plan their telemetry, calculate their orbit, and then move on to the next mssion, all were there celebrating the launch. Word comes through that the Delta team have put IRAS into the perfect orbit, within 10 metres of the planned 900 km (600 mile) altitude. Later that night several of us share a motel room. We see the launch on the television news. NASA, struggling with severe problems with hydrogen leaks in one of the Shuttle Challenger's main engines, has put out an excited press release about the wonders of IRAS. It is their first successful launch for some time. The protest at Vandenberg Air Force Base does not make it onto the news.

Later when I talk to my wife, Mary, she says that there was a lot of coverage on British television also. She says that it was even on John Craven's Newsround, a news programme for children on BBC television, so the boys saw it (my two stepsons, Adam and Jonathan, then aged 11 and 9). In his bulletin John Craven said that 'British experts had travelled out to Pasadena to help with the mission there'. A sort of fame.

IRAS refuses to cooperate

The next few days were frustrating and worrying. It appeared that there was a serious problem with IRAS. Each time it passed over the ground station at Chilton it responded to its commands to start a programme of tests, but when it returned it had always parked itself in a safety position, pointing directly away from the earth and with the instruments switched off. To try to understand what was happening NASA ordered their whole Deep Space Network of ground stations, normally used for tracking missions to other planets, to watch for the passage of IRAS and to interrogate its house-keeping data (information about every aspect of the spacecraft and telescope's operations, usually only transmitted to the ground twice per day). In this way NASA were able to catch IRAS in the act of disobeying its orders from the ground station and parking itself in safety mode. In a brilliant piece of detective work and repair, Richard van Holtz and his spacecraft team were able to discover that the problem lay in one of the sensors which recorded the direction of the sun. Its electronic box was being warmed up too much, generating electrical spikes, which the onboard computer read as a change in orientation of the spacecraft. The computer interpreted this as a danger to the telescope, which must never look near the sun or earth for fear of boiling away all the helium coolant, and immediately commanded the spacecraft into safety mode. Richard's team recreated the error in a simulation at Chilton, rewrote the ROM (read only memory) software to suppress the effects of the spikes and then transmitted the new ROM to the spacecraft computer. IRAS was back in business.

The first data were successfully shipped, via microwave links around the world, from Chilton to Pasadena. Although the cover was still in place on the top of the dewar, so there was as yet no view of the sky, electrical signals were still being read from the detectors, so we could pretend these were real observations and test our software. Morale at JPL was not high as everything seemed to be going wrong, nothing seemed to be working as expected. Tom Hibbard and I were therefore delighted to be able to report at the next morning meeting of the IRAS scientists and engineers that we had successfully run the data through our detection system after only one day. Something had worked, even if it was only the simplest part of the system.

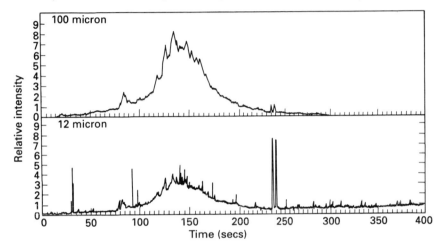

Fig. 5.4 First scans across the sky by IRAS, as seen in two of the four survey wavelength bands at 12 and 100 microns. The strong peak is the Milky Way and the spikes in the lower profile are stars.

After five days of tests, the environment of the spacecraft was deemed to have settled down enough to eject the cover from the dewar. In fact there was no choice, because the cover, which had its own supply of liquid helium coolant, was beginning to warm up. The scientists crowded excitedly into the office of the IRAS project manager, Gerry Smith, to hear over a radio link the command being given at Chilton. The cover blew off successfully and we could hear Fred Gillett's voice calling out 'I can see sources'. The characteristic rise and fall of the lines on the strip chart showed that the detectors were seeing astronomical sources. Later that day we received a fax of the first scan round the sky. The Milky Way and dozens of sources could be clearly seen. Our survey of the sky at far infrared wavelengths was beginning.

First view of the far infrared sky

6

THERE IS SOMETHING immensely exciting about seeing the sky for the first time in a new waveband. Could some completely new kind of object swim into view? In the 1970s my colleagues from Queen Mary and Westfield College (QMW), Peter Clegg and Peter Ade, and I had tried to get an impression of the sky at a wavelength of 1 millimetre using the National Radio Astronomy Observatory Millimeter Wave Telescope at Kitt Peak, Arizona. Unfortunately, our detectors were not sensitive enough to see much and though we looked towards dozens of sources representing the main known kind of extragalactic object, we managed to detect only a handful.

In the case of IRAS, there was no doubt that we would see many new sources. However, there was a long road from the billions of electrical signals being sent down by the satellite to anything like a picture. Following our first look at scans round the sky there now began a long and intensive period of several months during which we tuned our software and tried to make sure that we were finding all the sources that could be seen on the strip charts. A particular set of scans made very early in the mission, part of what came to be known as the 'Minisurvey', were analysed endlessly, making little changes in the programmes here, little improvements there.

Working at NASA's Jet Propulsion Laboratory

A key person in this process was Tom Chester, who headed the group of JPL software engineers responsible for making the IRAS Catalogues. A man of boundless cheerfulness, he developed an encyclopaedic knowledge of the system and all its problems and it was very enjoyable working with him to solve some of them. The bane of our lives was a NASA manager who was supposed to supervise our efforts. He understood little of what we were doing but he interfered endlessly. Later he was transferred to another NASA division and it was a source of quiet satisfaction to us that NASA did not think it necessary to replace him.

The NASA environment seemed to stimulate its employees, especially its managers, into a continual search for new jargon, especially acronyms, and for new metaphors. One individual used the phrase 'Hit me with that again' at almost every meeting (when he had not grasped the meaning of what had been said). For another we were always 'between a rock and a hard place'. The struggle for new metaphors led to some spectacular examples of mixed ones and John Fowler compiled a wonderful collection of these (and other felicitous phrases) from IRAS project meetings. I give here some of the best:

> *'You guys turned me around since I first walked into this door'*
> *(Feb 1977, JPL)*
> *'We're talking about something that's still being talked about'*
> *(Sept 1977, JPL)*
> *'Maybe we did confuse the water' (Nov 1977)*
> *'I doubled them by a factor of two' (Feb 1980, JPL)*
> *'I'm just fishing at straws' (Aug 1980)*
> *'Aren't you kind of touting your own horn' (March 1981, JPL)*
> *'We want to be sure it has a bucket to fall into semantics-wise'*
> *(Mar 1981)*
> *'Can we get together verbally?' (Mar 1981)*
> *'That really took the steam out of my sails' (Feb 1982, JPL)*

The clouds of Hibbard

For the first few days after the ejection of the cover from the IRAS dewar, we were still not exactly sure where the telescope was pointing.

Tom Hibbard and I were checking out the computer programmes which were supposed to scan the raw data from the infrared detectors and pick out the astronomical sources. One afternoon Tom brought me some plots of the detector output for an orbit from the previous day. For most of the orbit the detectors behaved normally, every so often responding to some astronomical source. Suddenly at one point all the detectors went haywire. 'What's this stuff, then?' Tom asked. Our first thought was that it must be some debris left over from the ejection of the satellite's cover a few days previously. However, I stayed late that night puzzling over the spaghetti-like jumble of signals. I found that the same thing had happened on several other orbits that day, and at about the same point on the orbit. Gradually I realized that the output from different orbits could be pieced together to make a map. The satellite was in fact passing over hundreds of very bright sources. When I went to *Norton's Star Atlas* to see roughly where the satellite was supposed to be pointing, the penny dropped. We had 'discovered' the Large Magellanic Cloud. I posted the scans on the wall of the corridor outside my office, with the label 'The Clouds of Hibbard'. Careful analysis of the satellite's pointing by another IRAS scientist, Eric Young, showed that the spaghetti was in fact the Tarantula Nebula, in the Large Magellanic Cloud.

The Tarantula Nebula, catalogued as the star 30 Doradus, is one of the largest star forming clouds known. It lies at a distance of about 160 thousand light years from us. Some astronomers have argued that it contains a star of over one thousand times the mass of the sun, more than ten times more massive than any star known in the Milky Way. However, it is more likely that the nebula is illuminated by a compact cluster of ten or twenty massive stars.

As they were in their usual difficulties about extracting funding from Congress, NASA were desperate to release some early results from IRAS and my cut-and-paste map was redrawn by JPL graphic artists in lurid blood-red colours for distribution to the world's media.

Close encounter with Comet IRAS-Iraki-Alcock

Soon after the discovery of the Clouds of Hibbard, IRAS made another, slightly more significant discovery, a new comet. A small group had been set up on NASA's initiative at the ground station at Chilton, consisting of two young British scientists, John Davies and Simon Green,

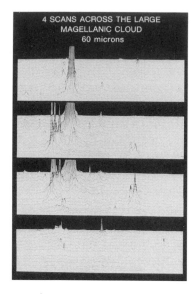

4 SCANS ACROSS THE LARGE
MAGELLANIC CLOUD
60 microns

Fig. 6.1 (a) The 'clouds of Hibbard'.

THE NEW YORK TIMES, TUESDAY, FEBRUARY 22, 1983

THE NEW YORK TIMES, TUESDAY, FEBRUARY 22, 1983

Orbiting Telescope Glimpses 'Birth' in Space

Tarantula

Area of Infrared Scan

Photo of Large
Magellanic Cloud,
as seen from Earth

By JOHN NOBLE WILFORD
In its first days of operation, a new
telescope orbiting the earth has returned infrared images showing

obtained in the years of looking
through instruments on the ground, in
balloons or high-altitude aircraft.
Infrared radiations are invisible to

cently spawned a cluster of massive
stars, each 10 to 100 times heavier
than the sun, or a single "monster"
star thousands of times more massive

(b) Newspaper headlines.

to study each day's data and look for objects moving across the sky. They were not interested in very fast-moving objects – these would be satellites passing overhead. But objects that moved significantly in a day compared with the 'fixed' stars definitely were interesting. These could be either asteroids or comets. IRAS was to find several comets, but this first one, discovered on April 26th, 1983, was especially exciting because it was the closest comet to earth for two hundred years. At 7 million kilometres, it passed at less than twenty times the distance of the moon. Its tail stretched out behind it some 400,000 kilometres. When the orbit of the comet was first calculated at Chilton it was not certain it would miss the earth. The best estimate was that it would

Fig. 6.2 (a) Comet IRAS–Iraki–Alcock.

SUNDAY CORRESPONDENT 29/7/90

LOS ANGELES TIMES

ASTRONOMY

Near miss that wasn't the end of the world

A BRITISH astronomer has revealed how scientists agonised over whether to issue a global doomsday warning in 1983, following the discovery of a comet – apparently heading

JPL spots comet closing in on Earth

An infrared telescope launched into space made its first major discovery yesterday. A comet that will fly only 3 million miles from home – a scientists announced yesterday.

That's the closest any comet has come to Earth in more than 200 years, according to Russ Walker, an astronomer with the Jet Propulsion Laboratory and part of an international team monitoring signals from IRAS, the Infrared Astronomical satellite.

The comet first flashed by IRAS last week, but scientists initially believed it was an asteroid. But later readings indicated the object was a comet "moving across the sky pretty fast."

(b) Newspaper headlines.

miss, but we could not be sure. We sat round at JPL digesting this fascinating information and debating whether we should warn the world's population of this worrying fact. Personally, I thought we should, because if the comet did hit, say, London or New York and we had failed to give a warning we were likely to be unpopular, to say the least. Typical scientific caution prevailed, however, and the IRAS team waited a few more days to get a better estimate of the orbit. The danger of a collision receded.

Because none of the IRAS scientists had discovered a comet before, nobody knew the correct form for announcing this kind of discovery. The key thing you have to do is to communicate the orbit to the Cometary Data Centre at Harvard. A new comet is then named after the discoverer. The IRAS team were so late in sending this information that our comet had already been sighted by two amateur astronomers, Iraki, from Japan, and George Alcock, from Britain. What should have been Comet IRAS, and legally should have become Comet Iraki–Alcock, became Comet IRAS–Iraki–Alcock.

Two weeks later the comet became visible to the naked eye as a faint blur of light and bulletins went on the television news programmes telling people where to look for it. Several of us made an expedition to a large amateur observatory in the mountains behind Los Angeles to get a sight of the comet. Once you knew where to look it was easy to find and when I got back down to La Crescenta, where I was staying, I could still see it even standing under a street light.

As the comet passed the earth, it was mapped with radio telescopes, so the size and mass of the cometary nucleus could be estimated. It turned out that we had been even luckier than I had imagined. The nucleus was 10 kilometres (7 miles) across, with a mass of a million million tons. Had it hit the earth, there would have been an immense catastrophe, comparable to the one believed to have put an end to the age of the dinosaurs. There is increasing evidence that the wave of extinctions 65 million years ago which wiped out one third of all species on earth, including the dominant species of the dinosaurs, was caused by the impact of an asteroid or comet with about the same mass as Comet IRAS–Iraki–Alcock. This would have dug a crater 20 kilometres (13 miles) across and spewed immense amounts of dust into the earth's atmosphere, sufficient to blot out the sun over much of the earth for many months. There is growing evidence that this asteroid of

65 million years ago in fact hit the earth in the ocean, in the Gulf of Mexico. Thus in addition to the climatic impact of the huge dust cloud, there was a tsunami or tidal wave a kilometre high, which was responsible for many of the extinctions that followed.

Comet IRAS should therefore give us pause for thought. Not only do we need to save the planet from our own damaging activities, we also need to think how we might survive this type of cosmic catastrophe, which we narrowly missed in April 1983. There will be a worrying moment for our descendants in the year 2126 when comet Swift–Tuttle, which was first discovered in 1862 and then crossed the earth's orbit again in 1992, returns. On that occasion it is predicted to cross the earth's orbit on July 11th, 2126, while the earth crosses the comet's orbit only one month later on August 14th. If the comet were delayed by a month on its 134-year orbit, a collision is possible. However, as the comet is travelling very fast, at 60–70 kilometres a second, it will take only three minutes to pass the earth, so our descendants would have to be very unlucky to suffer a collision on this occasion. Over the millennia, though, comet Swift–Tuttle will make many close passages to earth.

It was the last of the IRAS comets and asteroids, asteroid IRAS 1983TB, which established the link between comets and three other fascinating phenomena: *Apollo asteroids*, *meteorites* and *meteors*. Asteroids are lumps of rock ranging in size from one to hundreds of miles across and most of them are circling the sun in a ring between Mars and Jupiter, the Asteroid belt. However, the orbits of some, the Apollo asteroids, like some comets, cross the orbit of the earth. Luckily, most cross the earth's orbit when the earth is somewhere else on the orbit, but occasionally one does hit the earth and the consequences can be dramatic. Such collisions were much more common in the early days of the solar system and an impression of their effect can be seen in the form of the vast craters frozen into the moon's surface. The corresponding giant craters left on earth have been eroded away by wind and rain. As I have mentioned above, it may have been an Apollo asteroid which finished off the dinosaurs and many other species by throwing up a huge cloud of dust and turning day to night for months on end. What is the origin of these menacing cosmic wanderers?

Meteorites are lumps of rock, iron or coal-like (carbonaceous) aggregates which crash to earth from the sky. They have been venerated throughout history. The Kaaba, the sacred black stone at Mecca, the centre of Islam, is almost certainly a meteorite, and a similar stone was venerated by North American Indians. Yet it took astronomers many centuries to believe in the reality of stones from the sky. Even though thinkers from the Renaissance onwards recognized that celestial bodies were likely to be made of the same material as the earth, they still seem to have been trapped in some Platonic idea of the inviolability of the earth. In 1803 a commission of experts was sent by the French Academy of Sciences to the town of L'Aigle to investigate reports that thousands of stones had fallen on the town from the sky. They came away reluctantly convinced. It took that rarest of events, the disintegration of a meteorite over a populated area, to convince scientists of what had been known to many isolated individuals throughout history.

Meteors, on the other hand are particles of dust flashing through the earth's atmosphere, which heat up until they become incandescent and glow as a 'shooting star'. Meteors, or shooting stars, are seen especially frequently on particular nights of the year. I remember sleeping in the open on a summer night in France and seeing many hundreds, the famous Perseids. In 1866 Giovanni Schiaparelli realized that meteor streams occur when the earth passes through the orbit of a dying comet which has left its debris strewn along the way. Most of the impressive meteor streams are associated with known comets. For example, the August Perseids move in the same orbit as Comet Swift–Tuttle, the November Leonids follow that of Comet Tempel and the May Aquarids are associated with Comet Halley.

Asteroid 1983TB turned out to have the same orbit as the Geminid meteor stream. Its appearance was that of an Apollo asteroid, a dead lump of rock with no gas or dust tails. Thus at one blow it was demonstrated that at least some of the Apollo asteroids are simply the nuclei of dead comets which have exhausted their envelope of dust, gas and ice. Many meteorites too are almost certainly cometary debris. Detailed analysis of their composition shows that others are fragments from collisions between asteroids in the asteroid belt, and still others are fragments chipped off the moon or other planets by asteroids. For several comets, notably Comet Tempel, IRAS was able to map the trail of debris spread out along the orbit of the comet.

In January 1993, a group of astronomers, physicists and engineers met in Tucson, Arizona, to discuss 'Hazards due to Comets and

Asteroids'. They proposed a network of telescopes, 'Spaceguard', dedicated to discovering and tracking bodies on earth-crossing orbits. This seems an excellent proposal. Less reassuringly (in fact, positively terrifyingly), NASA has seriously explored the idea of hundreds of orbiting nuclear warheads as a defence against near-earth asteroids. This seems to be a case where the medicine is far more dangerous than the disease.

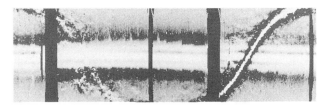

Fig. 6.3 The zodiacal dust bands discovered by IRAS, seen as a broad pair of horizontal bands across the sky close to the plane of ecliptic. The S-shaped band is emission from the Milky Way.

Yet another insight into the violent past of the solar system came with the IRAS team's discovery of bands of dust spread round the sky close to the plane of the *ecliptic*. The ecliptic is the plane of the earth's orbit around the sun and, approximately, of the orbits of the planets and asteroids. Frank Low had the idea of displaying a sequence of scans round the sky and these were plastered all over the corridors between our offices. On each scan the bands could be seen on either side of the peak due to the ecliptic plane. These bands turned out to be at least partly due to the debris of collisions between members of families of asteroids which move in similar orbits within the asteroid belt, occasionally colliding with each other and showering debris along their orbits. My research team and I have spent several years trying to find out whether some part of the bands might have a more distant origin, perhaps in some ring of dust in the outer solar system connected to the cometary cloud. So far we have been unable to settle this issue.

Life at the ground station in Chilton

While we were endlessly reprocessing the Minisurvey data at JPL, trying to tune up the software which was to make the IRAS catalogues, the survey continued to press on remorselessly. Peter Clegg, a colleague and friend at QMW who was in charge of the science operations at the ground-station at the Rutherford–Appleton Laboratory at Chilton,

Oxfordshire, has described for me a typical day at the ground station.

> *There were usually about thirty people at the control centre, divid-
> ed into three groups: the spacecraft operations team, mainly
> Dutch, the American telescope team and the British science opera-
> tions team. Richard van Holtz, the Mission Operations Manager,
> was in overall command and was responsible for the health and
> safety of the satellite. As Resident Astronomer, I was responsible
> for making sure that we achieved the scientific goals of the mis-
> sion, but I had tremendous support from a team made up of Phil
> Marsden, from Leeds University, and Jim Emerson, Stella Harris
> and Helen Walker from Queen Mary College.*
>
> *After the years we had spent preparing for the mission we were
> naturally excited and apprehensive during the launch. Although
> IRAS was launched from California, all commands after lift-off
> were sent from Chilton, either directly as the satellite passed over-
> head or via NASA's network of ground stations. We successfully
> sent the command to open the vital helium valve soon after lift-off
> but we did not manage to contact the satellite until the very last
> moment of its first pass over Chilton: it was an anxious few min-
> utes. All went well at the next pass ninety minutes later, and the
> spacecraft and telescope teams began checking that all was well
> with IRAS and its instruments. A week later we blew off the cover
> which had been sealing the vacuum and protecting the telescope
> during launch. As the cover drifted away, we saw the signals from
> the detectors drop dramatically as IRAS looked at the cold depths
> of space for the first time. Twelve hours later we looked at the first
> data and were amazed at how good they were. In fact it took me
> some time to convince myself that we were looking at real data
> and not re-analysing some of the synthetic data we had used for
> testing the processing chain!*
>
> *For the first few weeks, most of us were at the control centre for
> every twice-daily pass and for much of the rest of the time as well.
> On the 'prime' pass we had about twelve minutes to acquire the
> spacecraft, tell it to send down, or 'dump', the previous twelve
> hours' data from the tape-recorders on board and to send up the
> sequence of instructions for the next twelve hours. Sometimes we
> failed to complete the task during the prime pass and had to use*

the six-minute 'back-up' pass ninety minutes later. These occasions could be quite tense. Between passes we checked the data which had been dumped and prepared a schedule of extra scans to re-survey parts of the sky where the data were noisy or missing. We also planned the programme of observations of individual astronomical sources which were used to fill up parts of the orbit which were not needed for the survey.

After the first month or so, when operations became more routine – although IRAS was ever ready to provide excitement if we got too complacent – we tended to operate a shift system and even managed to get home for the odd weekend.

A planetary system in the making ?

During the early months of 1983 the Infrared Astronomical Satellite, IRAS, made a remarkable discovery about the bright star Vega. Because the star is very bright and extremely well-studied at optical wavelengths, it had been chosen as a calibration standard for the IRAS survey. We had assumed that we could simply extrapolate the visible and near infrared measurements to the far infrared wavelengths studied by IRAS. However, George Aummann and Fred Gillett, two American members of the IRAS team working on the calibration problem at the IRAS ground-station at Chilton found that there was something very unusual about Vega. While the other bright calibration stars behaved as predicted, Vega was much brighter than expected in the far infrared. The excess emission appeared to be coming from material at a temperature of 80 K ($-193°C$), compared with the temperature of the star's surface of 10,000 K. The inference was that there is a thin screen of solid particles ('dust') situated at about 80 times the sun-earth distance from Vega. The observations also allowed the deduction that the particles are relatively large compared with most samples of interstellar dust, at least 1 mm in diametre.

Now Vega is known to be a very stable star which does not vary its light output or eject any material from its surface. The dust shell surrounding Vega must have been there since the star was formed. The amount of material involved turns out to be comparable to that involved in the planets of the solar system. We seem to be seeing a planetary system in the making.

The IRAS scientists immediately checked out other stars similar

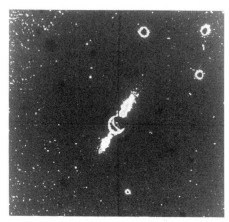

Fig. 6.4 Image of the dust shell around Beta Pictoris, made at optical wavelengths.

to Vega and found some other examples with dust shells, of which the best is Beta Pictoris. Ground-based observations by Bradford Smith and Richard Terrile at the Las Campanas observatory in Chile have shown that the dust round this star is distributed in the form of a disc, just as would be expected for a proto-planetary disc. They found that the disc has an overall diameter of 800 sun–earth distances but that the inner part of the disc, within 30 sun–earth distances of Beta Pictoris, seems to have been cleared of dust. This zone is comparable in size to the region occupied by the planets of the solar system and so perhaps the material within that radius has already formed into planets. However, the existence of this dust-free zone is still a matter of controversy and there is certainly no direct evidence for planets round Vega or Beta Pictoris.

Search for Planet X

In addition to our natural curiosity about other planetary systems, much interest has focused on whether there could be a tenth planet in our solar system. Only the five naked eye planets were known to the ancients. William Herschel discovered Uranus in 1781 during his telescopic survey of the northern sky, Neptune was discovered in 1846 from its perturbations to Uranus's orbit and Pluto was found by Clyde Tombaugh in 1930 during a programme designed to discover the cause of small unexplained perturbations to the orbits of Uranus and Neptune.

However, we now know that Pluto is too small to have caused these perturbations, which remain unexplained. The discrepancy is even more acute if observations of Neptune by Galileo in 1613 and by Lalande in 1795 were correctly recorded. Galileo was observing Jupiter at a time when Neptune was nearby and he appears to have recorded the position of Neptune, but not quite where it should have been if there were no tenth planet. Lalende was measuring the positions of stars and recorded a 'star' where no star should have been. In retrospect it is calculated that Neptune should have been close to the position recorded by Lalende, but again the position is not quite correct.

SUNDAY TELEGRAPH 3/11/91

Stargazers feel pulling power of Planet X

by Robert Matthews
Science Correspondent

BRITISH astronomers are this weekend investigating evidence for a new planet in our solar system, three times larger than the earth and orbiting the sun beyond the planet Neptune.

The evidence, uncovered by researchers at the Ruhr University in Bochum, Germany, will be discussed this week by astronomers at an international conference on the existence of new planets, being

sun as Neptune, which is currently the furthest planet in the solar system. Using the latest theories of how planets form, the researchers calculate that the mass of Planet X is about five times that of the Earth. If made of material common in the outer solar system, this would make the new planet almost 24,000 miles across, and the third largest in the solar system.

According to the calculations, Planet X takes about 760 years to complete one orbit. However, the orbit is very steeply angled relative

Fig. 6.5 Planet X headlines.

On the other hand, the tracking of the Pioneer and Voyager spacecraft on their long voyages to the outer planets of the solar system and beyond show that there can be no tenth planet in the plane of the ecliptic today. The orbit would have to be highly tilted to that of the orbit of the earth and other planets. Several astronomers still believe that Planet X does exist and are continuing to search for it.

There were two occasions in 1984 when I thought I had found Planet X in the data from the IRAS Infrared Astronomy Satellite. The first was when a bright cool source turned up on the IRAS map of the Galactic Centre region, lying right on the plane of the ecliptic. It soon turned out that the source was not moving across the sky, as a planet would, and its infrared spectrum showed that it was in fact a heavily dust-shrouded red giant star. However rumour somehow reached the magazine *New Scientist* and nothing could persuade them not to print the story 'IRAS discovers tenth planet'!

On the second occasion I was carrying out a more systematic search for cool, moving objects when to my amazement one turned up. Unfortunately, when a fellow IRAS scientist Russ Walker checked the position against the file of known comets and asteroids for me, the moving object turned out to be Comet Bowell, which had been discovered by chance a year or so previously out beyond Jupiter. Since then I have searched through the IRAS data for moving sources which might correspond to Planet X, without success. Although it is still not completely impossible that Planet X lies hidden somewhere in the IRAS data, I am now at least 70% certain that Planet X does not exist. This 70% represents the area of the sky in which I would certainly have found it: the other 30% includes the small area not surveyed by IRAS and the larger area of the Milky Way where it would be almost impossible to find. The discovery of Planet X would have been the crowning achievement of IRAS, but it was not to be.

Discovering the power of IRAS

FROM THE MOMENT when the US, Netherlands and Britain had decided in 1976 to collaborate in sending up a cryogenically cooled telescope, IRAS, to survey the skies at far infrared wavelengths, I had been convinced that we could use it for cosmological studies, to map the universe. When IRAS was originally designed we had hoped to see hundreds of thousands of sources, including tens of thousands of galaxies. But the performance of the detectors in tests on the ground had been so poor that immediately before launch I estimated we would see only 1000 sources, of which very few would lie outside our Galaxy. To our amazement and delight the performance of the telescope and detectors in orbit was so perfect that the original plans and hopes were easily exceeded. Altogether almost a million new infrared sources were discovered by IRAS.

It is striking that in the years before launch the name of IRAS was associated with crippling problems, exasperating and endless delays, and escalating costs. Once launched, the brilliant performance of IRAS made it an instant success. We knew immediately that there were wonderful discoveries waiting to be made in this unique survey. In fact our problem became the impatience of the astronomical community to know what these discoveries were and to see the data for themselves. Within the group responsible for generating the IRAS catalogues, we were reluctant to release the data until it had been properly calibrated and assessed. We knew that those who had in the past released prematurely catalogues of astronomical sources, containing a high proportion of spurious sources, had come to regret this bitterly. Although our astronomical colleagues cursed us at the time for the year or so they had to wait for the IRAS catalogues, I think the high quality of the catalogues justified the delay.

Fig. 7.1 The distribution on the sky of the infrared sources detected in the IRAS survey. The sources concentrated towards the plane of the Milky Way are stars and dust clouds where new stars are forming. Away from the Milky Way, most of the sources are distant galaxies.

The IRAS 'minisurvey' and the infrared cirrus

Within a fortnight of launch, the ground-station team at Chilton, led by Peter Clegg, had commanded the satellite through a thorough survey of a strip of the sky, the IRAS 'Minisurvey'. For months the science analysis team at JPL analysed the Minisurvey data, tuning the computer algorithms used to detect and confirm the reality of the sources of infrared radiation. Finally, in July 1983, we made the final adjustment to the hundreds of parameters involved in the preparation of the IRAS catalogues, and the generation of these catalogues began to roll. It was to take almost a year to complete, with one day's computing for each day's worth of data. The IRAS mission lasted ten months before the liquid helium coolant ran out, by which time it had surveyed 97% of the sky.

It turned out that the launch date of IRAS was exceptionally unfavourable for extragalactic studies and as a consequence we gained a completely wrong impression of the sky. The Minisurvey, carried out during the first fortnight of the mission so that we would have good data on one area of the sky and at least have something to publish if the helium ran out quickly, was in one of the worst parts of the sky we could have chosen, short of mapping along the Milky Way itself. It ran through the dense nearby molecular clouds in Orion, Taurus and Ophiuchus. There were thousands of cirrus sources and local star forming clouds. It was almost the most difficult area possible in which to tune our detection software. But in a way this may be why we ended up with such a marvellous catalogue. Having slaved for months to get the software to work on the terribly confused and complex Minisurvey area, the quality of the data in the high latitude sky, free of the confusing effects of cirrus and emission from star forming regions in our Galaxy, was a wonder to behold. We did not see these high Galactic latitudes until July of 1983.

(a) (b)

Fig. 7.2 (a) Tom Chester, who led the team of software engineers responsible for the preparation of the IRAS Catalogues. (b) Tom Soifer, who chaired the team of scientists working on the design and quality control of the IRAS Catalogues.

Of course the infrared 'cirrus' was itself one of the great discoveries of IRAS. The detection of infrared emission at wavelengths of 60 and 100 microns was not entirely unexpected. What was unexpected was how bright the cirrus clouds were at wavelengths of 12 and 25 microns. It became clear that some interstellar dust grains were at a much higher temperature than predicted. This turned out to fit in well with two other unexplained earlier discoveries. In 1973 Fred Gillett of Kitt Peak National Observatory, Arizona, and colleagues had found broad unidentified infrared features in the spectra of several sources. And in 1981, Kris Sellgren, as part of her thesis work, had found extended 5–20 micron emission around hot stars. At about the same time Steve Price was reporting from the AFGL rocket survey that the

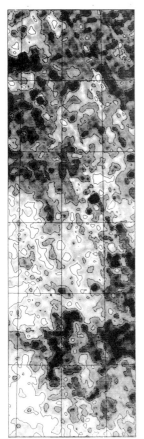

diffuse radiation from the Milky Way was brighter than expected at wavelengths of 4–20 microns. I remember him asking me what this could be due to and my replying, without much thought, that it was probably the integrated emission from circumstellar dust shells.

All these phenomena were in fact explained by very small grains, less than one thousandth of a micron in radius, or large molecules, in the interstellar medium. One candidate for this component is molecules of polycyclic aromatic hydrocarbons, which are common in the terrestrial environment (in tar or automobile exhausts for instance). A group led by the young French astronomer Jean-Loup Puget has been very active in pursuing this idea.

IRAS galaxies

Once we had correctly orientated the position of the telescope on the sky, following the detection of the 'Clouds of Hibbard', it became possible to identify many of the IRAS sources. It was clear rather quickly that while the 12 micron band was dominated by stars in our Galaxy, the 60 micron channel was seeing a quite different sky. Towards the plane of the Milky Way we were seeing emission mainly from dust clouds where new stars were being formed. Over the rest of the sky, though, we were seeing emission from thousands of galaxies. A few weeks after launch I compiled a list

Fig. 7.3 The infrared 'cirrus', far infrared emission from wispy filaments of interstellar dust.

of sources from the Minisurvey area we were using to tune our software algorithms, with tentative identifications for many of them. One of the Galactic sources was the dark cloud Barnard 3, which turned out to be the first IRAS 'proto-star', a star in the process of formation. There were many galaxies from existing optical catalogues and these formed the starting-point for our first study of IRAS galaxies, a study led by Tom Soifer of Caltech. The sources which did not have identifications in my list and which were not too close to the Milky Way formed the basis for the study led by Jim Houck of 'unidentified' sources. Most turned out to be more distant galaxies. By the summer of 1983 we had prepared a series of papers on the first discoveries by IRAS and submitted them to *The Astrophysical Journal*. They were published as a

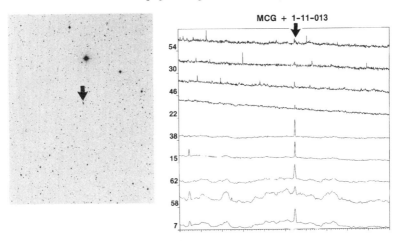

Fig. 7.4 (a) One of the first galaxies identified with an IRAS source.

THE TIMES   10/11/83

(b) Newspaper headline on the unidentified sources.

special issue of the Letters section of the journal on March 1st 1984 and press conferences were held simultaneously in the US, Netherlands and UK to announce the results, which stimulated several newspaper stories.

Meanwhile I had begun to study the sources found at high Galactic latitudes, which were free of the problem of contamination by sources in our own Galaxy. The four survey wavebands at 12, 25, 60 and 100 microns, would each be sensitive to material at a different range of temperature. Typically the 12 micron survey was sensitive to material at $300^{\circ}K$, while the 100 micron survey was sensitive to material at $30^{\circ}K$. Thus we would expect to see different types of astronomical object in each waveband. By studying how the numbers of sources changed as a function of brightness in each of the four IRAS wavelength bands, and by identifications of individual IRAS sources with stars and galaxies in pre-existing optical catalogues, it was becoming clear how sharply, away from the plane of the Milky Way, the sky changed from one filled with stars at a wavelength of 12 and 25 microns to one filled with galaxies at a wavelength of 60 microns. Moreover the small number of stars found at 60 microns could be easily recognized by their different infrared colours. At these high Galactic latitudes we also found that the quality of the survey was extremely high. The dream of using IRAS for cosmological studies was becoming a reality.

By comparing the positions of the infrared sources not identified as objects in optical catalogues with the Palomar and SERC Sky Survey photographic plates, it soon became clear that IRAS was picking up thousands of galaxies at the longer survey wavelengths of 60 and 100 microns. Even at high Galactic latitudes, though, emission from interstellar dust was being detected at 100 microns, the infrared 'cirrus'. It was important to understand this so that we could be sure that it was not dominating our galaxy studies. My group at Queen Mary and Westfield College has spent many years studying the cirrus and producing accurate maps of its distribution on the sky.

One of the studies we carried out was to compare the counts of the numbers of IRAS galaxies in different ranges of brightness, towards the north and south Galactic poles, respectively. Such a comparison depended on how accurately the brightnesses of faint sources were being measured. We were confident that the accuracy of the IRAS flux measurements was very good, but we had to argue our way through objections from many different directions. In the end I think we convinced most people that IRAS was indeed a remarkable cosmological tool. At 60 and 100 microns we found that there were more faint sources per unit area of a given brightness towards the north Galactic pole than towards the south. Since we seemed to be surveying to a distance of at least 150 million light years in each direction, this suggested that we are located in a density structure, albeit not of very great density contrast, of truly enormous dimensions, at least 300 million light years across. It is still not clear how common such structures are in the universe. Later we were to make a more statistical study of large-scale structure using the IRAS survey and this was to have dramatic consequences for ideas about how galaxies, clusters of galaxies and other large structures may have arisen in the universe (see Chapter 10).

The first systematic follow-up of the IRAS survey with ground-based telescopes

By the time we released the IRAS Point Source Catalog in November 1984, I was already planning a programme to study these IRAS galaxies from ground-based optical telescopes. I made an agreement with Tom Soifer that I would study one third of the sky around the North Galactic Polar cap (at Galactic latitude greater than 60°), while he and his American collaborators would study the remaining two-thirds. The point of this agreement was to avoid us doing the same work. I had persuaded Andy Lawrence to join QMW from the Royal Greenwich Observatory, and we set up a collaboration with an old friend of mine from student days, Michael Penston. With Michael's research student Kieron Leech and my research assistant David Walker, Andy measured the spectra of all the IRAS sources in our 844 square degree area around the North Galactic Pole and showed that, apart from a few stars, obvious from their infrared signature, over 99% of the 60 micron sources were galaxies, with recession velocities ranging from one 1,000 to 30,000 kilometres per second (up to 10% of the speed of light, or redshifts of 0.1). Using the Hubble law to convert the velocity into a distance, we could estimate the numbers of sources per unit volume in different ranges of luminosity. This quantity is called the *luminosity function* and is a key to studies of what the nature of the galaxies is. At the lower end of the luminosity range we were seeing infrared radiation from dust in most normal spiral galaxies. At intermediate luminosities we started to see galaxies in which a more than average amount of star formation is going on. Finally, at the very highest luminosities we were seeing galaxies undergoing huge bursts of star formation, which have become known as 'starburst' galaxies.

In the event Tom Soifer decided not to pursue the galaxies in the North Galactic Polar cap. Instead he and his Caltech team concentrated on a sample of bright IRAS galaxies, for which most of the velocities were already known. The luminosity function they derived from this sample agreed well with ours, as also did those derived from other redshift surveys carried out by George Rieke of the University of Arizona and collaborators, and by others.

I will return to the question of the nature of ultraluminous IRAS galaxies in Chapter 11. The most striking possibility that presented itself following our pilot survey of the north Galactic cap was that we could map the galaxy distribution over the whole sky, out to a depth far greater than any previous large-scale survey. One motive for doing this was to see whether we could find structures on even larger scales than the large clusters of galaxies already known. This would be crucial to understanding how structure arose in the universe. Another was to try to study the origin of our own Galaxy's motion through space, which had been detected through a small (0.1%) anisotropy of the microwave background radiation around the sky (see p.63). To understand how we know that our Galaxy is moving through the cosmological frame, we have to leave the IRAS story and go back almost twenty years to the story of the discovery of the microwave background and the subsequent decades of intensive study of this radiation. We shall see that there is a close link between the cosmological discoveries made with the IRAS survey, and those of the microwave background, culminating in the detection of 'ripples' in the microwave background by the COBE satellite (Chapter 12).

The microwave background story

THE DISCOVERY OF microwave background radiation by Arno Penzias and Robert Wilson at Bell Telephone Laboratories in 1965 was one of the most important and unexpected astronomical discoveries of the century. Although several cosmologists came close to making a clear prediction of the existence of this radiation, the discovery actually came as an accident and was made by two young radio-astronomers who had not worked in cosmology and did not at first appreciate the significance of what they had discovered. The importance of the discovery was that it showed that we live in a Big Bang universe, dominated in its early stages by radiation. At the time of the discovery this was only one of several possible models of the universe under consideration. In particular the steady state cosmology had many supporters at this time, particularly in the UK.

The microwave background radiation is intimately bound up with the IRAS story as we shall see in later chapters. Although the microwave background radiation is extremely uniform, or isotropic, round the sky, small departures from this isotropy eventually allowed the absolute motion of our Galaxy through space to be measured by several groups in 1978. The speed of our Galaxy's motion turned out to be surprisingly high. Almost a decade later my colleagues and I were able to explain the origin of this motion in terms of the attraction of a dozen clusters of galaxies within 300 million light years of us. We were then able to use our IRAS galaxy surveys to map the 'lumpiness' of the universe on large scales and hence probe how galaxies and clusters formed. This 'lumpiness' in the universe today turned out to be directly related to the 'ripples' in the microwave background radiation found by COBE in 1992. The combination of the IRAS and COBE studies of large-scale structure pin down rather precisely how structure must have formed in the early universe.

The triumph of the Big Bang theory

Although the cosmologists who had been working on the Big Bang model of the universe immediately accepted Penzias and Wilson's discovery as proof that their preferred model was correct, other cosmologists, particularly those who supported rival models, were slow to accept this view. One aspect of the Big Bang model which some scientists and philosophers found uncomfortable was that it predicts the universe has a finite age and has its origin in an initial explosive instant of creation. For some this seemed a bit too close to religious notions of creation and they had welcomed the ageless nature of the Steady State universe.

I think I can pin-point the moment at which the Big Bang model of the universe became *the* model accepted by most cosmologists. In 1973 a symposium was held in Cracow, Poland, to commemorate the 500th anniversary of the birth of Copernicus. The subject was 'Confrontation of Cosmological Theories with Observational Data'. The title of the meeting suggested that a range of cosmological models would be compared with astronomical observations. But in fact almost every paper at this conference seemed to show that the universe had the simplest possible form, in which an almost perfectly smooth and isotropic universe of the kind first considered by Einstein in 1917 starts from a state of infinite density and temperature, the Big Bang, spent its first few hundred thousand years in a 'fireball' phase dominated by radiation, and then cooled to the matter-dominated universe we observe today.

In his amusing introduction to the published proceedings of the conference, the great Russian physicist and cosmologist (and one of the key figures in the development of the Soviet hydrogen bomb) Yakov

Zeldovich set out what seemed to be a very open-minded philosophy of cosmology. We cannot simply solve backwards in time the equations governing the evolution of the universe, he said, but have to take arbitrarily chosen variants of the initial state and follow their evolution to the present and thus to a confrontation with observation. Zeldovich pointed out that the snag of this procedure is that it is dependent on the prejudices, likes and dislikes of authors and 'perhaps even their subconscious Freudian attitude to such things as order, chaos, antimatter'. Hence the importance of the confrontation with observation, in which 'false theory fades'. I liked the suggestion that cosmologists, by their choice of cosmological model, might reveal their unfortunate infant experiences and neuroses.

But was it really cosmological theories, in the plural, that were being confronted with observations? Apart from a few dissenting, sceptical and dissatisfied voices, only one theory was discussed at the conference. The gestalt switch to a new scientific paradigm, described by Thomas Kuhn in his influential book *The Structure of Scientific Revolutions*, had occurred and the isotropic Big Bang model had become the new paradigm. Within that framework the theoreticians set to work on the details: the problem of galaxy formation, the structure of the initial singularity, the cosmic abundances of the light elements. Thus Kuhn's sociological description of what he thinks happens in a scientific community when a scientific revolution occurs did seem to apply rather accurately to the establishment of the hot Big Bang theory.

In a review of the conference proceedings for the scientific magazine *Nature*, I drew attention to this shift in concensus amongst cosmologists and asked whether we would again have to wait the 2000 years between Ptolemy and Copernicus to see this picture overthrown. In response to this review I received a witty letter from Zeldovich which asked me what alternative theories I wanted to support, was it the by then discredited Steady State theory of Bondi, Gold and Hoyle, or was it the non-cosmological redshift theories of Arp and others? Already it was clear, in the most charming way possible, that those who are not for us are against us.

An idea which appeared for the first time at this conference, in a paper by an old student acquaintance of mine, Brandon Carter, was the *anthropic principle*, which argues that if the universe was not as it is, galaxies and stars could not have formed and we would not be here to observe it (see Chapter 3). As this sounded a bit like Aristotle's geocentric view of a universe with humanity at its centre, I sarcastically concluded my review: 'Come back Aristotle, all is forgiven.'

Zeldovich commented sardonically in his letter: 'Perhaps the situation in Cracow *was* Aristotelian. I take your challenge. But it seems to me that you are mourning for something like the astronomy of Assyro-Babylonian or Egyptian time, an occult science of priests, with astrological perversions.'

Penzias and Wilson accidentally discover the microwave background

The situation had been very different when I began research in cosmology in 1964. At that time the debate between the Big Bang and the Steady State theories, personified by their most public advocates Martin Ryle and Fred Hoyle, respectively, was still at its peak and seemed the most important issue of all to work on. But a year later a discovery was announced which changed everything. On May 13th, 1965 a modest and laconic letter entitled 'A measurement of excess antenna temperature at 4080 Mc/s' reached the *Astrophysical Journal*. The paper, only 600 words long, describes in a brief, almost brusque,

Fig. 8.1 Arno Penzias (right) and Robert Wilson, of AT&T Bell Laboratories, the discoverers of the microwave background radiation.

style the first project undertaken by two young radio-astronomers, Arno Penzias and Robert Wilson, since their move to Bell Telephone Laboratories. They had been working with the 20 feet horn reflector built by Bell Laboratories engineers for the Echo satellite project, in which a signal was bounced off the satellite and received on the other side of the Atlantic. Using this antenna, they had measured the brightness of the sky when the antenna was pointing in the zenith (vertical) direction. This brightness can be characterized by the *blackbody temperature* which would give the same brightness. I have mentioned previously that material which is a perfectly efficient absorber or emitter of radiation is described as black. A terrestrial realization of a blackbody is the interior of a cavity maintained at a very uniform temperature, like a furnace. The surfaces of the stars are, approximately, blackbodies. The German physicist and founder of quantum theory, Max Planck, showed that the spectrum of radiation emitted by a blackbody had a shape, peak wavelength and intensity which depended only on the temperature of the blackbody. The peak wavelength for material at 5800°K, the surface temperature of the sun, is at 0.5 microns, in the visible waveband; for 300°K (the approximate temperature of the human body) it is at 10 microns; for 30°K it is at 100 microns; and for 3°K it is at a

Fig. 8.2 The Bell Laboratories antenna with which the microwave background was discovered in 1965.

wavelength of 1 millimetre. In each case a blackbody gives a characteristic distribution of intensity at wavelengths around that peak wavelength. Any measured intensity or brightness at a particular wavelength
can be translated into an equivalent blackbody temperature which
would give the same brightness at that wavelength. Penzias and Wilson
had found a temperature of 6.7°K, of which 2.3°K was due to atmospheric absorption and 0.9°K was due to Ohmic losses in the antenna.
This left 3.5°K unaccounted for.

Because there was a possibility that this excess signal could be
due to a problem with the antenna itself, the two astronomers
described how they had taken great care to clean and align the joints
between the three sections of the antenna to reduce the losses in the
structure, how they tested for leakage and the antenna horn seams for
loss (by taping up all the seams with aluminium tape). The cleaning
involved removing what Penzias has referred to as 'a white dielectric
substance' left by a pair of nesting pigeons. The pigeons, too, had to be
removed. First they were transported to another Bell Laboratories site
some distance away, but when they returned immediately to the antenna, they were, as Penzias later put it, 'discouraged by more active
means'. Penzias and Wilson checked the 'backlobe' response of the tele-

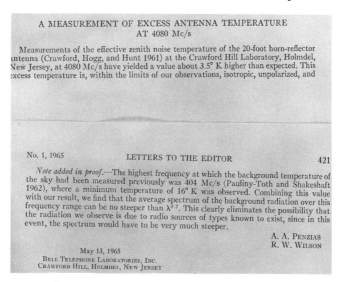

A MEASUREMENT OF EXCESS ANTENNA TEMPERATURE
AT 4080 Mc/s

Measurements of the effective zenith noise temperature of the 20-foot horn-reflector
antenna (Crawford, Hogg, and Hunt 1961) at the Crawford Hill Laboratory, Holmdel,
New Jersey, at 4080 Mc/s have yielded a value about 3.5° K higher than expected. This
excess temperature is, within the limits of our observations, isotropic, unpolarized, and

No. 1, 1965　　　　　LETTERS TO THE EDITOR　　　　　421

Note added in proof.—The highest frequency at which the background temperature of
the sky had been measured previously was 404 Mc/s (Pauliny-Toth and Shakeshaft
1962), where a minimum temperature of 16° K was observed. Combining this value
with our result, we find that the average spectrum of the background radiation over this
frequency range can be no steeper than $\lambda^{0.7}$. This clearly eliminates the possibility that
the radiation we observe is due to radio sources of types known to exist, since in this
event, the spectrum would have to be very much steeper.

A. A. PENZIAS
R. W. WILSON

May 13, 1965
BELL TELEPHONE LABORATORIES, INC.
CRAWFORD HILL, HOLMDEL, NEW JERSEY

*Fig. 8.3 The paper by Penzias and Wilson announcing the discovery of the
microwave background radiation.*

scope to ground radiation by beaming a small transmitter at the back of the telescope. They even used a helicopter to beam a known source of microwave radiation at the telescope to test its response. They noted that 'this excess temperature is, within the limits of our observations isotropic, unpolarized, and free from seasonal variation'.

Had this paper appeared in the *Bell System Technical Journal* like other work with the horn antenna, astronomers might hardly have noticed it, as they hardly noticed Karl Jansky's earlier (1933) discovery at Bell Laboratories of radio emission from the Milky Way, described in Chapter 2. For Penzias and Wilson did not themselves immediately recognize the significance of their discovery. However, they had the good fortune to be near Princeton, where the group led by Bob Dicke and Jim Peebles was only too aware of the significance of a 3°K microwave background. This was just what would be expected in a hot Big Bang universe which is initially dominated by radiation. Such a model of the universe had been advocated by George Gamow, originally from Russia, in the 1940s. In 1924, as a student, Gamow had attended and been inspired by Alexandr Friedman's lectures on General Relativistic cosmological models, given shortly before Friedman's tragic early death.

In early measurements with the radiometer which he invented, Bob Dicke had noted in 1946 that the temperature of any cosmic background radiation was less than 20°K. He had then turned to other experiments and to the development of the Brans–Dicke cosmology, a rival theory to General Relativity which has subsequently been effectively disproved. By 1965 Dicke, in collaboration with Jim Peebles at Princeton, had reinvented or at least revived a version of the Hot Big Bang model of George Gamov, and was predicting the existence of cosmic microwave background radiation with a blackbody spectrum (in seminars if not yet in print) and had two young researchers, P. G. Roll and David Wilkinson, building a system to measure the temperature of this background. In March 1965 Peebles submitted one paper to *The Astrophysical Journal* on the effect of cosmic blackbody radiation on galaxy formation, and another to the journal *Physical Review* entitled 'Cosmology, cosmic blackbody radiation, and the cosmic helium abundance'. The plans and dreams of the Princeton group must have been rudely shattered by the news of Penzias and Wilson's discovery. Peebles, Dicke, Roll and Wilkinson hastily wrote an elegant paper explaining the significance of the radiation as the relic of the 'primeval fireball' phase of the hot Big Bang, and outlined the physics of this radiation-dominated, opaque phase of the universe's history. This paper appears immediately preceding the one by Penzias and Wilson, with the far more dramatic title 'Cosmic blackbody radiation'. The paper Peebles had already submitted on galaxy formation had to be revised to include the Penzias and Wilson background temperature, but one sentence from the abstract which had been sadly overtaken by events failed to get revised: 'There is good reason to expect the presence of blackbody radiation in an evolutionary cosmology, and it may be possible to observe such radiation directly.'

Penzias and Wilson included only one very cautious remark on the significance of their 3 K background: 'A possible explanation for the observed excess noise temperature is the one given by Dicke, Peebles, Roll and Wilkinson in a companion letter in this issue.' Penzias had heard about the Princeton group's ideas and invited them over to the Bell Laboratories station at Holmdel. Without that visit the only reference to the excess noise might have been the one that appears in a foot-note to another paper by Penzias and Wilson, on radio emission from the strong radio source Cassiopeia A, received in its original form by *The Astrophysical Journal* on April 1st, 1965, a month before their paper announcing the discovery of the background radiation: 'The equivalent temperature of the antenna when pointed at the zenith is about $7°K$, $2.3°K$ of which is due to absorption by oxygen in the atmosphere.'

While it is probably inevitable that Dicke and his group would have discovered the cosmic blackbody radiation within a year or so, it was also inevitable that Bell Laboratories would sooner or later identify any source of background noise in the microwave region, as earlier they had, through Jansky, identified the effect of the Milky Way in the radio region. It is taking nothing away from Penzias and Wilson to say that part of the credit for their discovery must go to the thorough, professional and single-minded work of the Bell Laboratories scientists and engineers over the years in developing sensitive receivers and reliable antennae. As Penzias and Wilson note in their historic paper, the $3°K$ excess is implicit in earlier work by Bell Laboratories scientists (Degrasse, Hogg, Ohm and Scovil) with the same antenna. The excess noise was a well-known problem at Bell Laboratories, but it took the careful work of Penzias and Wilson to show that it was not noise in the antenna itself. Interestingly there is a very similar antenna at Plumeur-Boudou in Brittany, France, which I came across when on holiday there. This was the other end of the Echo experiment and is now part of a science exhibition. French radio-astronomers may have missed an opportunity there to discover the microwave background themselves!

In fact there was a French radio-astronomer who may have come close to anticipating Penzias and Wilson. E. Le Roux was doing some work in 1956 on mapping emission from the Milky Way and arrived at an estimate that the temperature of any background emission was less than 3°K. If he was correct in his estimates of the accuracy of his measurements, he must have been close to detecting the microwave background. Even closer was the Ukrainian radio-astronomer Tigran Shmaonov, who made measurements at microwave frequencies in 1955 with a horn receiver. He found excess radiation with a temperature of about 3°K. The Russian radioastronomer Youri Pariskij told me recently that Shmaonov had consulted him about whether such radiation could come from the Milky Way. Quite correctly, Youri told him that it could not and so Shmaonov decided he must have underestimated his errors. He announced a background temperature of 3.5°K with an uncertainty of 3°K either way, in other words not a significant detection. Another measurement which in retrospect can be seen to be a detection of the microwave background was made by the Canadian, A. McKellar at the Dominion Observatory in 1941. Studying absorption lines in the optical spectra of stars due to the interstellar cyanogen radical (CN), he noticed that this was often found in an excited state with a temperature corresponding to about 2.3°K. This result was quite well known in fact. In his classic textbook on molecular spectra, Herzberg comments 'this temperature has of course only a very restricted meaning'. Fred Hoyle recalls that he used McKellar's result to argue to Gamow that the Hot Big Bang model, for which Gamow and his collaborators were in the 1940s predicting a background temperature of 5°K, was wrong.

The Nobel prize for Penzias and Wilson

In 1978 Penzias and Wilson were awarded the Nobel Prize for Physics. I was spending a year at the University of California at Berkeley at the time and I was startled to be rung up by the scientific journal *Nature* and asked to write an editorial about Penzias and Wilson. They gave me just twenty-four hours to do this. I rushed around looking up all their scientific papers and unravelling the story I have described above. I was able to talk to Charles Townes, himself a Nobel Laureate and Arno Penzias's supervisor, about his memories of Arno. In my article I also referred to Penzias and Wilson's pioneering work in molecular line astronomy (they discovered the famous 2.6 millimetre line of carbon monoxide and many other important interstellar molecules). Twenty-four hours later I travelled up the hill to the Space Sciences Group and painstakingly typed out the telex to go to *Nature*. Later I was delighted to get a letter from Penzias saying that the *Nature* editorial had 'the place of honor on our family bulletin board (the door of our refrigerator).'

I was curious to know what had motivated them to track down the noise in the antenna so ruthlessly and asked Arno Penzias about this. In addition to the absolute calibration of Cassiopiea A, they had also been trying to measure the spectrum of the Milky Way at around this time (this had been part of Penzias's PhD project). Penzias replied: 'Confusing as it sounds, we were sorting out the noise in our antenna while making flux measurements of Cassiopeia A in order to measure the background from the Galaxy. How's that for a compound sentence?'

One curious feature of the story of the discovery of the cosmic blackbody radiation was the apparent ignorance of the scientific literature of all concerned. Neither the Bell Laboratories nor the Princeton workers referred to Gamow's idea of a hot Big Bang, published in *Nature* in 1948, or to the prediction of Gamow's American collaborators Ralph Alpher and Robert Herman, published in 1949, of a radiation temperature at the present epoch of $5°K$.

Gamow's original idea was that all the elements of the periodic table would have been made in the Big Bang, but it soon became clear that most elements are in fact made much later in the history of the universe, in the interior of stars. The latter idea was already being pur-

---

Fig. 8.4 (a) Nuclear reactions during the Big Bang. Most protons and neutrons end up as hydrogen or helium, with traces of deuterium (^2H), helium-3 and lithium, (b) the predicted abundances of the light elements according to the Big Bang model, compared to those observed. The broken vertical lines show the allowed range of mean density, the two solid vertical lines the critical density (see p. 127) for two values of the Hubble constant.

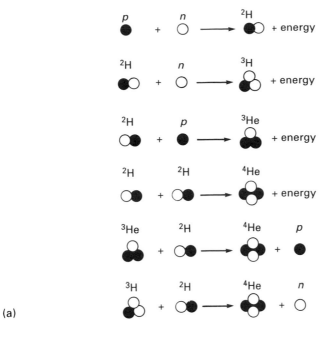

(a)

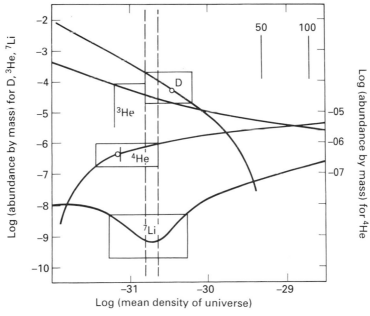

(b)

sued by Fred Hoyle in the 1940s. In 1957, Margaret and Geoff Burbidge, Willie Fowler and Fred Hoyle finally showed conclusively that all the elements from carbon onwards are made in the interior of stars or during supernova explosions. Only for the light elements deuterium (heavy hydrogen), helium, lithium, beryllium, and boron was the abundance relative to hydrogen obviously too great to be made in stars. In 1966, following the discovery of the microwave background by Penzias and Wilson, Jim Peebles showed that the correct abundance of helium would be produced in the Big Bang. Later, in 1974, Bob Wagoner, working with Willie Fowler and Fred Hoyle, showed that the hot Big Bang also predicted the correct amounts of deuterium and lithium. The French cosmologists Hubert Reeves and Jean Audouze, working with Willie Fowler and David Schramm, had shown in 1972 that beryllium and boron are made by the action of cosmic rays, atomic nuclei travelling through our Galaxy close to the speed of light, on helium atoms in interstellar space, a process now known as 'spallation'.

After the *Nature* editorial appeared I received an indignant letter from Ralph Alpher and Robert Herman saying that they did not believe the Bell Laboratories and Princeton workers did not know of the work of Gamow and themselves. As evidence, they sent me a copy of the paper Peebles had submitted to *Physical Review* in March 1965, which refers to a 1953 review by themselves. They also thought I should have given them more credit for their prediction that the background temperature was 5°K. My feeling had been that although Alpher and Herman had calculated a radiation temperature of 5°K, they did not seem to have appreciated that this meant that there would be background radiation at microwave wavelengths.

Another calculation of the helium production in a hot Big Bang which failed to make the connection with microwave background radiation was published by Fred Hoyle and Roger Tayler in 1964. The scientists who came closest to a clear prediction of the discovery were in fact the Russians A.G. Doroshkevich and Igor Novikov, in a paper published in 1963. They understood clearly that the implication of a hot Big Bang model was microwave background radiation, and they looked up the Bell Laboratories Technical Journals to check what the limits on such a background were from the Echo antenna. Unfortunately they misunderstood the papers of Ohm and other Bell Laboratories scientists, and thought that the observed noise in the antenna was entirely accounted for by atmospheric radiation. In the same year the eminent Russian theoretical physicist Yakov Zeldovich, hitherto a supporter of Gamow's hot Big Bang idea, launched an attack

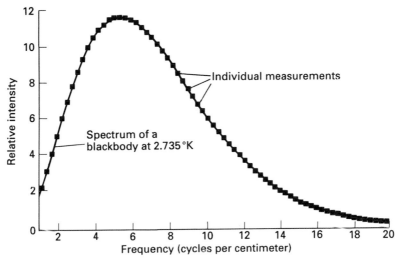

Fig. 8.5 (a) The spectrum of the microwave background radiation measured by the Cosmic Background Explorer, or COBE satellite in 1989. There is an extremely good fit to a Planck blackbody spectrum.

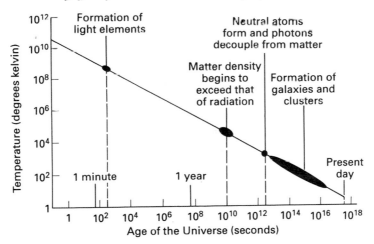

(b) The time evolution of the average temperature of the universe from a time one second after the Big Bang until the present day. As the universe expands, its average temperature and density both fall steadily.

on the theory as incompatible with observations and turned to consider the possibility of a cold Big Bang, with no radiation. He appeared to think that the hot Big Bang's prediction of a helium abundance of about 10% by number (25% by mass) was inconsistent with observations, possibly confusing the two different ways of quoting the abundance.

Over the next few years the radiation was almost always found to give the same temperature, about 2.7–3.0°K, whatever wavelength was observed. The spectrum was that of a blackbody, matter in perfect thermal equilibrium with radiation. Actually there was a period, in the late 1970s and early 1980s when it looked as if a distortion from the perfect blackbody form had been detected at submillimetre wavelengths, by groups first at Berkeley and then in Japan. Joe Silk and John Negroponte, of the University of California, Berkeley, and I speculated that a pregalactic generation of stars could have been responsible for generating the excess radiation. However the first results published by the Cosmic Microwave Background Explorer satellite (COBE, see Chapter 12) in 1990 showed that the spectrum of the background is a perfect 2.7°K blackbody, with no distortion detectable at all. And during the 1970s the degree to which the measured intensity of radiation looked the same in every direction turned out to be phenomenal. First the accuracy was 10%, then 1%, then 0.1%. Suddenly Einstein's apparently absurd 1917 oversimplification of an isotropic universe (p.23) began to look like a truly inspired guess.

Let us recap on the history of the universe implied by the discovery of the microwave background radiation. The universe today is filled with both matter and radiation, but the radiation today makes up only a tiny fraction of the total energy density (less than 0.1%). In the early stages, however, the energy density of the radiation was much greater than that of the matter. This is because, while the density of matter scales with the volume, or the cube of the size of the universe, the density of radiation scales as the fourth power of the size of the universe. Thus in the early universe, radiation dominates and the matter is of little significance. The temperature of the radiation decreases as the universe expands. Eventually, three hundred thousand years after the Big Bang, the temperature drops to 3000°K. At this temperature protons and electrons combine together to make neutral atomic hydrogen, and the universe becomes transparent to radiation. The ordinary baryonic matter (now consisting of a mixture of 75% neutral atomic hydrogen and 25% helium) and the radiation decouple from each other. The radiation continues to cool as the universe expands, retaining its black-body spectrum, until it reaches today's temperature of 2.7°K. Most of the photons of this relic radiation have wavelengths between 300 microns and 10 centimetres today (when they last interacted with matter, three hundred thousand years after the Big Bang, they were visible or near infrared photons). The microwave background radiation we detect with our telecopes today gives us a picture of the smooth, isotropic universe as it was only three hundred thousand years after the Big Bang.

Aether drift detected at last

Just before Penzias and Wilson received the Nobel Prize, the first deviations from perfect isotropy in the background were found. Groups at Berkeley, Princeton and Florence led by George Smoot, David Wilkinson and Francesco Melchiorri, respectively, all found that there was a smooth large-scale departure from isotropy across the sky. At a level of one part in a thousand, the temperature was slightly higher, and the sky slightly brighter, in one direction in the sky and slightly cooler in the opposite direction. For intermediate directions the temperature varied smoothly from the peak to the trough values. This became known as the *dipole anisotropy* in the microwave background radiation. The most natural interpretation is that the earth is moving

Fig. 8.6 George Smoot (foreground right) and the U2 aircraft with which the Berkeley group mapped the microwave background radiation.

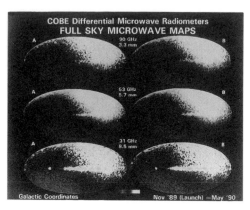

Fig. 8.7 The microwave background 'dipole' as measured by COBE in 1992. The microwave background radiation is slightly hotter and brighter to the upper right and slightly colder and fainter in the opposite direction, with a smooth variation between the two extremes at inter-mediate directions. This 'dipole' is believed to be due to the motion of our Galaxy through space.

with respect to the frame of reference in which the radiation looks isotropic. In the direction towards which the earth is moving, the radiation is then blueshifted, shifting the radiation to higher frequencies. The photons therefore carry more energy and the intensity of the radiation appears brighter. Similarly, in the opposite direction the radiation is redshifted to lower frequencies, so the intensity is reduced.

When allowance was made for the earth's motion round the sun and the sun's motion round the Galaxy, the net result was that our Galaxy was moving through space at a speed of 600 kilometres per second. This seemed an alarmingly high speed. There is a curious aspect to this. The famous Michelson–Morley experiment of 1885 showed that the speed of light in a vacuum is the same whatever the speed of the observer. This destroyed the concept of the aether, which had dominated nineteenth century physics, and launched the way for the Special Theory of Relativity in which all observers in uniform motion with respect to each other are equivalent. In cosmology, however, the situation is different. There appears to be an absolute cosmological frame of rest in which the universe looks perfectly isotropic. We are not in that frame so we see the microwave background dipole. In writing about the discovery of the dipole for *Nature* in 1977, I ironically headlined my piece 'Aether drift detected at last'. For almost a decade controversy raged about the origin of this microwave background dipole, and of our motion through space. It took the results from the IRAS survey, and its follow-up using ground-based optical telescopes, to get to the bottom of it.

The IRAS dipole and the 'death' of the Great Attractor

I HAVE DESCRIBED how in 1985 Andy Lawrence and I, with collaborators from the Royal Greenwich Observatory, carried out the first redshift survey of IRAS galaxies in a substantial area of the north Galactic polar cap. This was enough to demonstrate the immense cosmological power of the IRAS survey. At this point we conceived the idea of an 'all-sky IRAS galaxy redshift survey', in which we would measure the redshifts, and hence distances, of a large sample of IRAS galaxies spread all round the sky. This would have two goals. The first would be to study the origin of our Galaxy's motion with respect to the microwave background. The second would be to look for structure on scales larger than the largest clusters of galaxies known.

I learned that Marc Davis of the University of California at Berkeley was also thinking of doing a large IRAS galaxy redshift survey and I talked with him about the possibility of our doing a combined project. However, he was determined to carry out his own project, so we agreed to go our separate ways, remaining in friendly rivalry ever since.

The origin of our Galaxy's motion through space

Our Galaxy's motion through space manifests itself in a 'dipole'

125

anisotropy of the microwave background radiation. The radiation looks slightly hotter and brighter in one direction on the sky, the direction of our motion, and slightly cooler and dimmer in the opposite direction. For directions in between there is a smooth change of temperature across the sky. In a galaxy survey like the IRAS 60 micron survey, the number of galaxies per square degree of a certain brightness varies across the sky in a much more irregular way, corresponding to the effects of different clusters of galaxies in the volume we are surveying. However, we can still try to find the direction on the sky which corresponds best to a dipole pattern of variation across the sky, with higher average galaxy density in that direction and lower average galaxy density in the opposite direction. The all-sky galaxy sample which I had extracted from the IRAS 60 micron survey turned out to be ideal for this type of study. For the first time we had a sample of galaxies which covered most of the sky, reached to great depths in space, had well-calibrated and accurate brightnesses (albeit at the unfamiliar wavelength of 60 microns) and was free from the effects of extinction by interstellar dust in the Milky Way.

The first study of this kind was carried out in 1986 by Amos Yahil, of the University of New York at Stony Brook, my research assistant, David Walker, and I, during a visit by Amos to Queen Mary and Westfield College. We analysed the IRAS galaxy distribution to find the direction which gave the best fit to a dipole variation in galaxy density. We found that the direction of the IRAS galaxy dipole agreed remarkably well with that of the microwave background dipole. The interpretation of this is that in the IRAS survey we are seeing the clusters of galaxies whose net gravitational attraction is responsible for our Galaxy's motion. Each cluster of galaxies pulls us towards it. The strength of the attraction we feel depends on the mass of the cluster and its distance. Using data from the IRAS galaxy redshift survey carried out by QMW and RGO in the north galactic cap, Amos, David and I were able to show that the galaxies and clusters responsible for our motion probably lie within 300 million light years of us.

We were also able to make a rough estimate of the average density of the universe, because the speed with which our Galaxy moves depends on how much matter there is in the universe pulling it. Interestingly, this average density came out to be much higher than the amount of matter believed to be present in the universe in the form of normal 'baryonic' matter, that is, neutrons and protons, the stuff our bodies, the earth and the stars are made of. The density of baryonic matter can be estimated from the abundances of the light elements deuterium (heavy hydrogen) and helium. The higher the density of baryons in the universe, the more helium and the less deuterium were made during the Big Bang. Although some helium is made by nuclear fusion from hydrogen in the interior of stars, most was made during the Big Bang. Deuterium is not made in stars at all, but some of the deuterium made in the Big Bang is destroyed in the interior of stars, and this has to be allowed for.

The cosmological density parameter, Ω

The expansion of the universe is slowed down by the gravitational attraction of the matter in the universe. If the average density of the universe is low, the attraction is insufficient to halt the expansion and it continues indefinitely. On the other hand, if the average density of the universe is high, the expansion is eventually halted and the universe starts to collapse together again, ending in a second phase of infinite density, the Big Crunch. There is therefore a critical density which distinguishes models which will keep on expanding for ever from those in which the expansion will be halted and reversed. Cosmologists define a parameter Ω (*omega*), the cosmological density parameter, which is the ratio of the actual average density of the universe today to the critical density. If we could determine Ω, we would know the fate of the universe.

The critical density also determines the spatial curvature of the universe. For a density greater than the critical density ($\Omega > 1$), the universe has positive spatial curvature. This means that the angles of a triangle add up to more than $180°$, and a circle of radius r has a circumference less than $2\pi r$ and the universe closes up into a ball as we look out further and further. For a density less than the critical density ($\Omega < 1$), the universe has a negative spatial curvature: the angles of a triangle add up to less than $180°$ and circles of radius r have a circumference greater than $2\pi r$. If the density is equal to the critical density ($\Omega = 1$), the universe is spatially flat, as in Euclidean geometry.

All this presupposes that gravity is the only force acting on the universe. Einstein in fact introduced an additional force into the field equations of General Relativity, which becomes important on large scales, the 'cosmological repulsion'. If this force is present in the universe, then to

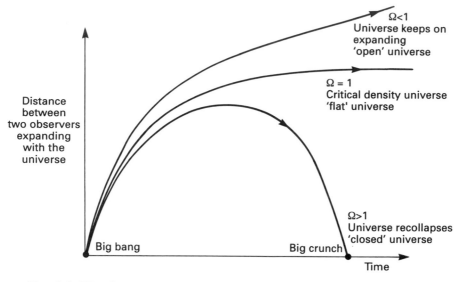

Fig. 9.1 The three possible fates of the universe: the closed universe recollapses to a Big Crunch, the open universe keeps on expanding for ever. Between the two extremes is the flat universe, in which the expansion slows to a halt at an infinite time in the future. Which universe we inhabit is determined by the average density of the universe.

determine the future of the universe we need to know both the average density of the universe and the magnitude of the cosmological repulsion.

One consequence of a cosmological repulsion is that the age of the universe can be much longer than the Hubble expansion time (the age is always less than the Hubble time if there is only gravity acting). The present estimate of the age of the oldest stars in our Galaxy is 13 billion years, so the universe must be at least as old as this. Now a Hubble constant of 80 km/s/Mpc corresponds to a Hubble expansion time of 13 billion years. Thus if the Hubble constant turns out to be greater than this value, which would imply a Hubble time less than 13 billion years, then there *has* to be a cosmological repulsion in the universe (unless there is something wrong with the stellar evolution calculations which are used to estimate the ages of the oldest stars in our Galaxy).

My PhD supervisor Bill McCrea pointed out in 1965 that the meaning of the cosmological repulsion was that it corresponded to the energy density of the vacuum. In classical physics it is therefore natural to assume that the cosmological repulsion is zero. This is not obvious, however, in modern particle physics theory, where the vacuum is assumed to be teeming with activity. Inflation theory, which I described in Chapter 4, takes advantage of this to postulate a phase in the early universe when the energy density of the vacuum, and hence the cosmological repulsion, was very high (see also Chapter 10).

Assuming for the moment that the cosmological repulsion is zero in the universe at the present epoch, the cosmological density parameter, Ω, becomes the key quantity to measure. The abundance of the light elements implies that ordinary baryonic matter contributes only about 3% of the critical density, i.e. $\Omega = 0.03$. If we add up all the starlight in galaxies, and work out the mass of the stars making this light, we find that visible stars contribute only 1% of the critical density ($\Omega = 0.01$). This shows that most of the ordinary baryonic matter is not in the form of visible stars. There are two ways that we can try to measure the total mass of galaxies, including luminous and non-luminous matter. One is to study the orbits of stars and gas clouds round the outer parts of the galaxies. The balancing of centrifugal force and gravity then tells us the mass inside that orbit. The other is to study pairs and groups of galaxies in orbit around each other. Again the orbital speed gives an estimate of the mass in the system. Both methods confirm that galaxies have halos of non-luminous, 'dark' matter. Adding up the total masses of galaxies determined in this way to get an estimate of the mean density of the universe, a value of about 10% of the critical value ($\Omega = 0.1$) is found. This is certainly enough to account for the missing baryonic matter.

What about matter between the galaxies? In the 1930s, the Swiss astronomer Fritz Zwicky had argued that giant clusters of galaxies, like those in the directions of the constellations of Virgo and Coma, had more matter in them than could be accounted for just by the galaxies. In the 1960s, Allan Sandage and others had hoped to estimate the total density of the universe in all forms of matter by studying how the brightness of galaxies changed with redshift. By selecting the brightest galaxies in clusters, a very clean 'Hubble diagram' of apparent brightness versus redshift could be made. For the nearby galaxies, the brightness simply varies as the inverse square of distance. At greater distances, this behaviour is modified by the effects of General Relativity,

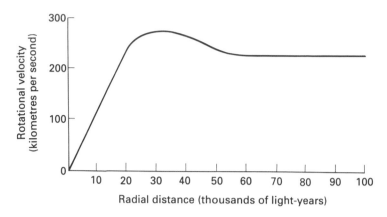

Fig. 9.2 (a) Rotation curve of the spiral galaxy M31. The rotational velocity does not fall off with radius in the outer regions as expected if there were only the visible matter in the galaxy disc. The almost constant value of the rotational velocity implies that there is a halo of dark matter surrounding the disc.

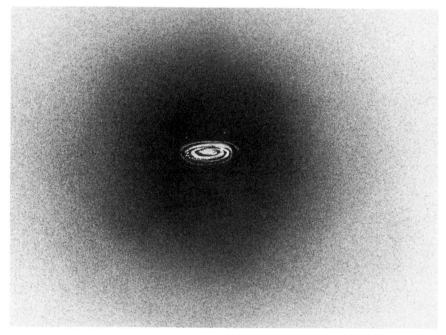

(b) Artist's impression of a dark halo surrounding a spiral galaxy.

and the modification depends on the density of matter in the universe. Hence, in principle, the average density of the universe could be determined. Unfortunately a young New Zealand astronomer, Beatrice Tinsley, showed in her PhD thesis in 1968 that the effects of the evolution of stars in galaxies were too strong to make this test worthwhile in the optical waveband. More recently, Simon Lilley and Malcolm Longair have tried to apply this kind of test to near infrared measurements of radio-galaxies. Their hope is that the effects of stellar evolution are not nearly so marked at infrared wavelength, but no really convincing answer for the mean density of the universe has emerged from this method yet.

The only hope for determining the total density of matter in the universe therefore lay with large-scale dynamical studies, which compare the speeds with which galaxies are moving around with the spatial distribution of the clusters of galaxies which generate these motions. Up to 1986, when Amos Yahil, David Walker and I published our preliminary result from the IRAS survey, such studies had generally given values for the average density of matter of around 10–20% of the critical density ($\Omega = 0.1$–0.2), more or less consistent with the density of ordinary baryonic matter derived from the primordial abundances of helium and deuterium. Our result was that the average density of the universe was about 70% of the critical value ($\Omega = 0.7$), with an uncertainty that would be consistent with the universe having the critical density ($\Omega = 1$). This was a very interesting result because for the first time it suggested that the density of the universe might be close to the critical value, and by implication that most of the matter in the universe is in some dark, non-baryonic form. This was clearly worth exploring further.

Birth of the QDOT all-sky IRAS galaxy redshift survey

Andy and I decided to draw in additional collaborators, Richard Ellis and Carlos Frenk from Durham, and George Efstathiou and Nick Kaiser from Cambridge, together with my two research students John Crawford and Will Saunders. I was rather pleased with the acronym QCD (for QMW–Cambridge–Durham) for the collaboration, because this is the acronym for the basic theory of particle physics called quantum chromodynamics, a well-known acronym to physicists. Unfortunately, when Nick moved to Toronto and George to Oxford, we had to change it to the less elegant QDOT. The first thing we had to decide was how big a survey to try to carry out. Nick Kaiser had shown earlier that if you

(a)   (b)   (c)

(d)

Fig. 9.3 Members of the QDOT (QMW–Durham–Oxford–Toronto) team,
who joined a collaboration to measure redshifts of an all-sky sample of
IRAS galaxies with the author and Andy Lawrence: (a) Carlos Frenk,
(b) George Efstathiou, (c) Nick Kaiser, (d) Will Saunders.
Opposite (e) Richard Ellis (left) and John Crawford (at the William
Herschel Telescope at the moment they completed the northern part of the
QDOT survey), (f) Andy Lawrence, at the Isaac Newton Telescope with the
Faint Object Spectrograph in the background.

want to study structure on a certain scale, the most economic way to do
this is with a sparsely sampled survey, where you take 1 source in n at
random from your catalogue and measure the redshift. He and George
Efstathiou calculated what the maximum value of n should be to achieve
the goal of measuring structure on scales of 100 million light years or
so. They came up with the answer that for the IRAS 60 micron survey
with its 13,000 galaxies (after excluding areas of the sky close to the
Milky Way), n should not be bigger than 10. I then made a guess at

(e)

(f)

how much telescope time I thought we might be able to get over the next few years, and guessed that we could do slightly better than 1 in 10. I opted for a 1 in 6 survey and this is what we designed and applied for.

The observing time for this survey was won in a gradual and piecemeal way from the committees which allocate time on optical ground-based telescopes. We got off to a good start with an observing run at the Isaac Newton Telescope on La Palma in the Canary Islands by Andy Lawrence and Nick Kaiser in June 1986, but our second week of time in January 1987 was wiped out by appalling weather. Andy Lawrence and I were the observers and we were at the summit on La Palma during the worst week of weather in living memory. Thirty inches of rain, close to the *annual* average on La Palma, fell in that week. The road from the Residencia, where the astronomers sleep and eat, to the summit, where the telescopes are, was strewn with rocks and even metre-sized boulders. Torrents of water poured down this road at several points. Wind speeds were persistently in the range 40–80 miles per hour. Bits of the roof of the Residencia were torn off in several places and there were leaks everywhere. Despite these conditions we travelled to the summit every night and refused to give up for the night until we were within an hour or so of dawn. During these long nights waiting for the weather to change, I conceived and sketched out my book *Universe*, so the week was not totally wasted. We were not able to open the telescope dome even once that week.

First scientific run with the William Herschel telescope

By July 1987 when I organized the 3rd International IRAS Conference at QMW, we had completed only 25% of the survey. Marc Davis and Michael Strauss, of Berkeley, were already presenting preliminary results from an all-sky survey of brighter IRAS sources. I was in despair. Then by a great stroke of luck we were successful in getting one of the first blocks of scientific observing time on the UK's new 4.2 metre William Herschel Telescope on La Palma, a four-week commissioning slot in December 1987. We were warned that the time might all have to be used for fixing the telescope. In the event the telescope worked almost perfectly. It is incredibly impressive to type in the co-ordinates of a galaxy position and see the field loom up on the guider television screen a few seconds later, knowing the vast piece of machinery which has swung round the sky to achieve this. Since we work in a control room separated from the telescope itself, we do not actually see the telescope moving.

The plan was to measure more than 50 redshifts a night, and we divided up into roughly four teams of two, each working for about a week. The first two teams, George Efstathiou with Carlos Frenk and then Nick Kaiser, were unlucky with the weather. Andy Lawrence and I also began with frustrating nights of bad weather. Then, when 12 of our 28 nights had passed with almost no redshifts measured, Richard Ellis took over from Andy, the weather cleared and we set off at a maniacal rate. On our fifth night Richard and I measured 104 redshifts, almost certainly a record, and we maintained something like this pace to the end of our shift. It was sobering to recall that the pioneers of observational cosmology like Edwin Hubble had to spend the whole night measuring the redshift of a single galaxy. Now we were measuring galaxy redshifts at the rate of one every few minutes. We also took some spectra of a new supernova which had just exploded. When Andy Lawrence, first with John Crawford and then with Will Saunders, took over from us in the following shifts, they continued at this pace, though with some problems from the brightening moon. Andy enlivened my Christmas by ringing up to report the discovery of a quasar with an unusual-looking spectrum. Eventually this was the subject of the first submitted paper based on data from the William Herschel Telescope. By the end of the run we had measured redshifts for 1100 IRAS galaxies and completed 80% of our survey. Two runs in the southern hemisphere on the Anglo-Australian Telescope at Siding Springs, and one further run with the William Herschel Telescope on La Palma, allowed us to complete the survey the following year. The long task of data reduction began, much of it carried out by Xia Xiaoyang of Beijing University, then studying for her PhD at Durham.

When we had the complete set of galaxy redshifts, we first studied the distribution of infrared luminosities of galaxies. While most galaxies have infrared luminosities similar to that of our own Galaxy, amounting to about 20–30% of their total energy output, there is a very interesting minority of exceptionally luminous infrared galaxies. In some of these, over 90% of their total energy is emitted at far infrared wavelengths. We shall return to these ultraluminous infrared galaxies in Chapter 11. We were also interested to see whether the probability that a galaxy has strong infrared emission depends on the age of the galaxy. As we look

Fig. 9.4 Clusters of galaxies, (a) Virgo,

(b) Centaurus.

out into the universe, we are also looking back in time, so we hoped to find out something about the history of the galaxy population. From the QDOT sample we found that in the past, galaxies tended to be stronger infrared emitters than they are today. Finally, we needed to measure the density of IRAS galaxies of different infrared luminosities accurately so that we could use them to map out the total density of matter.

A density map of galaxies within 500 million light years

Because our galaxies had been selected at infrared wavelengths, we were free of the problems of extinction by interstellar dust spread through our Galaxy and could map over a much larger fraction of the sky than any previous optical survey. The dust grains, typically one hundredth to one tenth of a micron in radius, absorb radiation strongly at visible wavelengths, but hardly at all in the far infrared. The sensitivity of IRAS also meant that we were probing deeper than any previous optical galaxy survey covering a substantial area of sky. Thus we were able to make a unique, three-dimensional map of the density of matter in a substantial volume of the local universe, of radius 500 million light years. The only drawback was that IRAS was sensitive only to 'spiral' galaxies, in which star formation is still taking place today. 'Elliptical' galaxies, which have for the most part formed all their gas into stars, and which comprise 20% of all galaxy types, were hardly seen at all by IRAS. This was to lead to lively debate subsequenty about whether the far infrared or visible pictures of the sky was the correct one.

Using the IRAS galaxy density map we were able to estimate the gravitational pull exerted on our Galaxy by the galaxies and clusters in our vicinity. The direction of the net pull on us agreed well with the direction our Galaxy is actually moving with respect to the microwave background radiation. We had therefore fully explained the dipole anisotropy of the microwave background, the only significant deviation this radiation shows from a perfectly smooth, isotropic distribution on the sky. A similar result had been found by Marc Davis and his collaborators from their survey of brighter IRAS sources, but we were able to show that a significant part of the gravitational pull on our Galaxy comes from clusters beyond the limit of their survey.

The 'death' of the Great Attractor

The IRAS survey gave us the first map of the galaxy distribution over

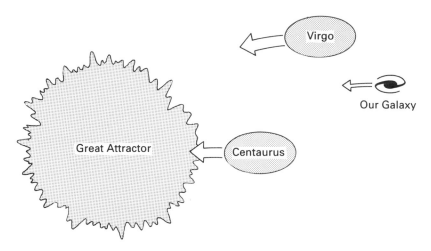

Fig.9.5 The 'Great Attractor' model for the origin of our Galaxy's motion through space.

the whole sky to a depth of 500 million light years. However, there had been several earlier attempts to look for large-scale structures in the distribution of galaxies, through the effects of such structure on the motions of galaxies. Around any large density structure, a large cluster of galaxies, say, we expect to see nearby galaxies falling in towards the cluster. Now from earth we can not measure the three-dimensional velocity of a distant galaxy (its speed and its direction), we can only measure the component of its velocity towards us or away from us in our line of sight to the galaxy. Imagine that it is night-time and we hear a train moving away from us, whistling. Provided we know the true pitch of the whistle, then from the pitch we hear we can estimate the speed of the train along the line from us to the train. If the train is moving radially away from us then that will be the true speed of the train, but if the direction of the train's motion is inclined to the line joining us to the train, then the estimated speed will be lower than the true speed. If the train is moving perpendicularly to the line joining us to the train, we would just hear the true pitch of the whistle, because the train's velocity does not have any component on the line joining us to the train.

The recession velocity of the galaxy in the line of sight, measured by its redshift, is a combination of the recession velocity of the galaxy due to the expansion of the universe, which is always radially away from us, and the component in our line of sight of the motion the galaxy has due to the pull of other galaxies and clusters, called its 'peculiar motion'. This is not because there is anything strange about it, but only because it is the motion peculiar or particular to that galaxy. If all we know is the redshift of a galaxy, we cannot tell how much of this implied recession velocity is due to the expansion of the universe and how much is due to the peculiar motion. But if we can directly measure the distance to the galaxy, then we can infer what the expansion velocity should be (making some assumption about the Hubble constant – see Chapter 4) and hence deduce the peculiar velocity along the line of sight. Such an analysis has been carried out for quite large numbers of galaxies using two distance methods I described in Chapter 4, the Tully–Fisher distance method for spiral galaxies (based on an observed correlation between the rotation velocity of a galaxy and its luminosity) and the 'D-σ' distance method for elliptical galaxies (based on a correlation between the size of the galaxy and the average random velocity in the galaxy). Once we have a map of peculiar velocities for galaxies at different distances and in different directions on the sky, we can begin to see whether these velocities are what we would expect from the clusters of galaxies that we already know about in our locality.

The nearest cluster of galaxies to us is the Virgo cluster, about 60 million light years away. This was first recognized by William Herschel during his telescopic 'sweeps' of the sky at the end of the eighteenth century. Today we know of several thousand galaxies which belong to this cluster and we recognize that the Local Group of Galaxies, of which the Milky Way galaxy and M31, the Andromeda Nebula, are the two dominant members, lies on the outskirts of this great cluster. Other prominent clusters of galaxies within 150 million light years include the Hydra, Centaurus and Fornax clusters. All these very prominent clusters are named after the constellation in which they are mainly located on the sky. Obviously, though, there is no connection between the stars which make up the constellation, which are relatively nearby stars in our Galaxy, and the very much more distant galaxy clusters.

When the motion of our Galaxy with respect to the microwave background radiation was discovered through the microwave background 'dipole' (see Chapter 8), it was immediately apparent that the direction of this motion was not simply towards the Virgo cluster. The Local Group of galaxies is not just falling towards the nearby Virgo cluster. An early suggestion by Gustav Tammann and Alan Sandage in 1985 was that our motion was due to the combined effects of the Virgo, Hydra and Centaurus galaxies. The direction of our motion is indeed somewhere between the direction of these three clusters.

A rather different picture emerged from a study of a large sample of several hundred elliptical galaxies by a group who allowed themselves to be called the Seven Samurai (Donald Lynden-Bell, Alan Dressler, David Burstein, Sandy Faber, Roger Davies, Roberto Terlevich and David Wegner). They had invented a new distance method for elliptical galaxies, the D-σ method, and seemed to find very large peculiar velocities, over a thousand kilometres per second, away from us in the vicinity of the Centaurus cluster. They inferred that there must be a very large structure behind Centaurus responsible for these motions. As a lesser consequence, this huge object would also be primarily responsible for our motion. This huge object was dubbed 'The Great Attractor' by Alan Dressler and became the object of great scientific and media interest. For a while there were several conferences every year with titles like 'large-scale structure and large-scale streaming motions', at which the problem of the huge peculiar velocities inferred in Centaurus and elsewhere would be aired.

I was sceptical of these anomalous streaming motions from the start. At a small conference held at the old Royal Greenwich Observatory at Herstmonceux, Sussex, during a blizzard in 1987, before the 'Seven Samurai' had published their results, I argued that the new distance method, though very interesting, was not accurate enough to make these claims. I also said, there and on many subsequent occasions, that the IRAS surveys did not show any sign of a large object behind Centaurus.

The IRAS density maps from the QDOT survey, which we had used to explain our motion with respect to the microwave background, confirmed that there was no sign of a very large density structure behind the Centaurus cluster and therefore cast doubts on the claims that the Centaurus cluster is 'streaming' away from us at 1000 km/sec. Our results were published in December 1990 and attracted coverage in newspapers and on radio and television. The inventors of the 'Great Attractor' were not pleased by the strong terms in which I cast doubt

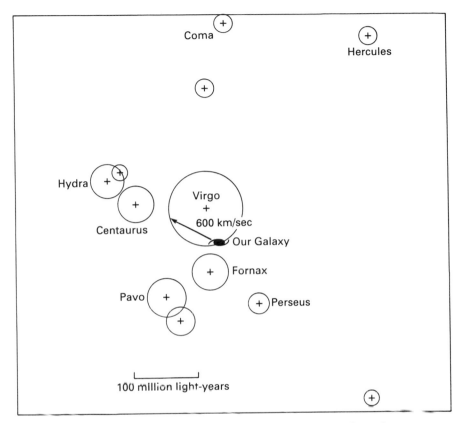

Fig. 9.6 The IRAS 'dipole' and the origin of our motion through space. The location of 12 clusters of galaxies within 500 million light years whose combined gravitational attraction accounts for the motion of our Galaxy through space. The size of the circle denotes the magnitude of the pull on our Galaxy. The horizontal direction (to the right) is towards the centre of the Milky Way and the vertical towards the north Galactic pole.

on its existence. Their main complaint was that the version of the 'Great Attractor' model I had criticized was no longer the one they believed, so

> *That is not what I meant at all.*
> *That is not it, at all.*

Subsequent studies of the Centaurus region have come to con-

THE INDEPENDENT ON SUNDAY   7/10/90

Universe loses its great attraction

ASTRONOMERS believe they have solved the puzzle of the Great Attractor, which was supposedly the biggest object in the universe but had never been seen. The solution is that it does not exist.

For several years, scientists

By **Steve Connor**
Science Correspondent

This radiation is slightly hotter and brighter in one part of the sky and cooler and dimmer in the opposite direction. This implies that

Fig. 9.7 Headline on the 'death' of the Great Attractor.

flicting conclusions about this streaming motion. Galaxy maps have shown that the clusters in this region are very extended and that there is a new normal-sized cluster behind Centaurus, but they have not yet revealed anything capable of causing the streaming. The picture derived from our IRAS maps is that our Galaxy is being pulled by a dozen or so clusters, with the bulk of the net pull being due to the Virgo, Hydra and Centaurus clusters, as had been proposed by Tammann and Sandage.

Our study gave another very important result. If our Galaxy's motion is indeed due to matter distributed through the universe like the galaxies and clusters sampled by IRAS, then we could measure how much of this matter there must be. We confirmed the result found previously by Amos Yahil, David Walker and I, that the mean density of the universe is high, close to the critical density which distinguishes a universe which will keep on expanding for ever from one which will eventually fall back on itself into a final 'Big Crunch'. Marc Davis and his research student Michael Strauss had found the same result in their study of a brighter sample of IRAS galaxies. Because the density of matter in the form of normal baryonic matter is known to be much lower than this, our result meant that the universe has to be mainly filled with some form of invisible, exotic matter. What could this dark matter be?

Crisis for cold dark matter?

In January 1991 the QDOT team published a research paper in the journal *Nature* that, according to the headline writers of the world's newspapers, plunged cosmology into a crisis. Our paper described the results of a systematic study using ground-based telescopes of galaxies first found in a survey at infrared wavelengths by the Infrared Astronomical Satellite (IRAS). We had made the first deep, reliable three-dimensional maps of the distribution of matter in the universe. The results were dramatic: a serious discrepancy with the current standard model of how the universe of galaxies formed and evolved, a model which relies on an exotic, invisible form of matter, *cold dark matter*, to explain how galaxies and clusters of galaxies formed.

Dark matter in the universe

For 10 years we have known that there is more to the Universe than meets the eye. Astronomers have come to the conclusion that the Universe can not just contain the normal matter that our bodies, the Earth, the stars are made of. The bulk of this normal matter, the matter that we see through our telescopes, is in the form of baryons, that is, protons and neutrons. Other particles, such as electrons, contribute little to the mass of the Universe.

Not long after the moment of the Big Bang, baryons somehow clumped together to start making galaxies and clusters of galaxies. But at the very time this process should have been well under way, we find that the Universe was in fact incredibly smooth and uniform. We know this from observations of the microwave background radiation, believed to be the faint relic of the fireball phase of the Big Bang. The remarkable smoothness of this radiation shows that some 300,000 years after the Big Bang the (normal, baryonic) matter and radiation were uniform to at least one part in ten thousand. From such a smooth state there is

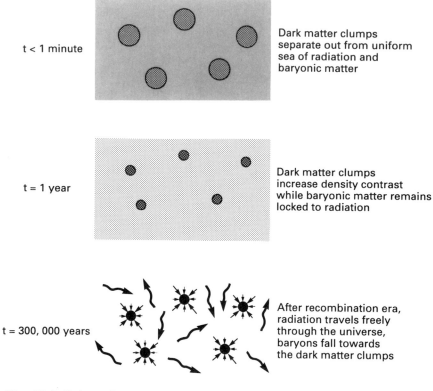

t < 1 minute — Dark matter clumps separate out from uniform sea of radiation and baryonic matter

t = 1 year — Dark matter clumps increase density contrast while baryonic matter remains locked to radiation

t = 300, 000 years — After recombination era, radiation travels freely through the universe, baryons fall towards the dark matter clumps

Fig. 10.1 Galaxy formation in a universe with dark matter. The dark matter decouples from the radiation at early times and density fluctuations in the dark matter begin to grow. When the baryonic matter decouples from the radiation it falls towards the dark matter lumps to form the visible parts of galaxies.

simply not time for gravity to assemble galaxies and clusters by today. The explanation astronomers favour is that the Universe must also contain matter in some so far undetectable dark form.

In the very early universe all forms of matter and radiation interact strongly with each other and share their energy equally. We say they are strongly coupled together. As time goes by, different components of matter separate out from the radiation, or 'decouple', and are then able to start to condense together under the action of gravity. The weakly interacting particles which make up the dark matter would have decoupled from the radiation at much earlier epochs than the ordinary baryonic matter, and any irregularities in their density would have had plenty of time to grow more pronounced under the action of gravity. By the time the baryonic matter decoupled from the radiation, the dark matter irregularities would already be well developed and would act as seeds for the condensation of the baryonic matter.

Theorists came up with two contrasting possibilities for this dark matter. One was 'hot' dark matter consisting of particles moving at speeds close to the speed of light in the early universe. The other type of exotic particle, 'cold' dark matter, has low velocities when it decouples from the radiation. Both types of dark matter interact only very weakly with normal baryonic matter, and so they are difficult to detect in terrestrial experiments. In the late 1970s, when theorists first aired the idea of dark matter, there was an immediate candidate for the hot variety in the form of neutrinos. Particle physicists usually regard neutrinos as having no mass and always moving at precisely the speed of light. However, the possibility that they have a small mass, in which case they would have to move at slightly less than the speed of light, has not been ruled out. There are no known particles with the right properties to be cold dark matter, but particle physicists predict several possible candidates. The two most popular, which fit in with ideas of how the forces of nature are unified together, are known as the axion and the photino.

Hot and cold dark matter act as seeds for galaxy formation in quite different ways. In a universe filled with hot dark matter only very large structures can form, because the fast-moving neutrinos stream out of and dissipate any smaller structure. These very large density irregularities then collapse to form pancake-shaped structures, the ordinary, baryonic matter collapsing with the hot dark matter into a sheet of gas, where shocks would rapidly cool down the gas. Out of these huge pan-

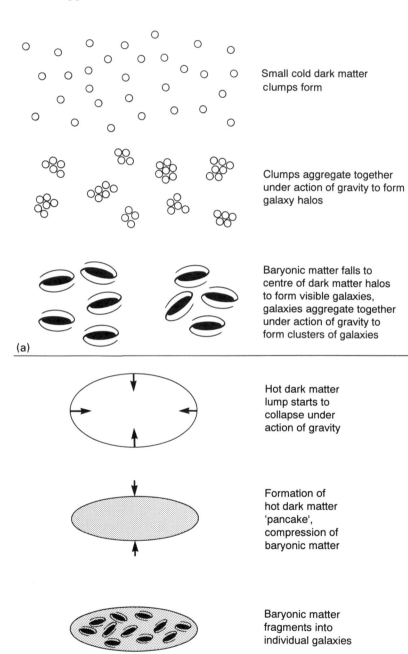

Small cold dark matter
clumps form

Clumps aggregate together
under action of gravity to form
galaxy halos

Baryonic matter falls to
centre of dark matter halos
to form visible galaxies,
galaxies aggregate together
under action of gravity to
form clusters of galaxies

(a)

Hot dark matter
lump starts to
collapse under
action of gravity

Formation of
hot dark matter
'pancake',
compression of
baryonic matter

Baryonic matter
fragments into
individual galaxies

(b)

(c)   (d)

Fig. 10.2 (a) 'Bottom-up' and (b) 'Top-down' scenarios for galaxy forma-tion. In the bottom-up scenario, the smallest perturbations collapse together first, and these then cluster together to make larger structures. In the top-down scenario, it is the largest structures that collapse first, and form huge pancake-shaped structures, out of which galaxies then condense. Above, (c) Jim Peebles, of Princeton University, (d) the Russian cosmologist Yakov Zeldovich.

cakes, smaller, galaxy-sized clumps would then condense. This is the 'top-down' picture for the formation of galaxies and clusters. In this picture we expect clusters to form before galaxies, and we also expect to find very large pancake-shaped superclusters. This picture of galaxy formation is associated with the Moscow group of the late Yakov Zeldovich.

In a universe filled with cold dark matter, on the other hand, much smaller clumps can form because of the slower velocities of the particles, and the smallest-scale structures tend to collapse together first. These small clumps of cold dark matter then coalesce together to act as seeds for galaxy formation. The normal, baryonic matter feels the gravitational attraction of these dark matter 'seeds' and falls towards them to make galaxies. Galaxies would then aggregate into clusters and larger structures under the action of gravity. This is the 'bottom-up' scenario. This type of galaxy formation picture, in which structure builds up from smaller scales to ever larger scales under the action of gravity, was particularly developed by Jim Peebles, of Princeton University, in the context of a universe dominated by ordinary baryonic matter.

One of the key tests of models for galaxy formation is how the probability of galaxy clustering changes with scale. If we have a galaxy A, what is the chance that a second galaxy B will be found at a specified distance away? During the 1970s, with a succession of collaborators, Jim Peebles used a series of different galaxy samples to study how this galaxy clustering probability changes with scale. He found that there was a smooth decrease in the probability from small to large scales, with no one scale of clustering being preferred to any other. Clusters of galaxies do not have a particular size.

The distribution of the galaxy clustering probability with scale at the present epoch, which is known as the *covariance function* for galaxies, is related to an important ingredient of Big Bang cosmology, the initial distribution of density fluctuations on different scales. Cosmologists hope to find that that this distribution of initial density fluctuations will be very simple, so that the formation of galaxies and clusters does not seem to be too much of an artefact. A particularly simple form for this distribution was suggested independently by Ted Harrison in 1970 and Yakov Zeldovich in 1972. Their idea was that as the horizon of the universe grows to a particular size, the average percentage deviation of the density of the universe from the mean on that scale would always be the same (the deviation would need to be about 0.01% to make galaxies and clusters by today).

Because it was difficult to match the top-down picture to the actual distribution of galaxies in clusters, it was the cold dark matter scenario which was pursued more vigorously. Computer simulations of the ways galaxies form and cluster together were explored by Marc Davis, George Efstathiou, Carlos Frenk and Simon White. Even with the aid of cold dark matter, they could not at first match the observed galaxy distribution, though. They found they had to assume that when galaxies formed, they tended to be concentrated, or 'biased', towards the places where there are exceptionally large fluctuations in the density of dark matter. Otherwise too bland and featureless a galaxy distribution resulted. Nick Kaiser had been the first to suggest this idea of biased galaxy formation.

The successes of the cold dark matter theory

Not only was this cold dark matter scenario able to match the statistics of the clustering of galaxies, but the simulations also looked very similar to the observed galaxy distribution. But was there any evidence that the universe actually contained dark matter? Radio and optical studies of the speed at which spiral galaxies rotate in their outer parts suggested that they are surrounded by halos of dark material. The orbital speed of a star round a galaxy gives an indication of the mass within the star's orbit, because the gravity acting on the star, which depends on the mass, has to be balanced by centrifugal force. The mass estimated this way always tended to keep on increasing with the distance from the centre of the galaxy, even when the distribution of stars was obviously petering out.

Analysis of the motions of pairs and groups of galaxies confirmed this result: galaxies are heavier than the total amount of visible, radiating matter would suggest. But when the masses of all galaxies, including their dark halos, are added together to give an average density for the universe, the total comes to only about 10% of the 'critical' value. A universe with less than the critical density will expand for ever. In a universe of density higher than the critical value, gravity will ultimately halt the expansion and the universe will recollapse to a 'Big Crunch'. Theorists tend to favour the universe having almost exactly the critical density, partly because it avoids our existing at a rather special epoch at which gravity and the expansion are still nearly in balance, but mainly because it is a requirement of a particularly successful theory of the early universe – the inflationary model – which we encountered in Chapter 4. The inflation of the universe leaves it very close to being perfectly flat spatially and, as we saw earlier (p.127), this means the density of the universe is extremely close to the critical value at the end of the period of inflation. Even today we would still expect the mean density to be within a small fraction of a percent of the criticial value.

Alan Guth of the Massachusets Institute of Technology proposed inflation, in 1981, as a phase that occurred just after the Big Bang, when the whole of the observable universe inflated exponentially from an infinitesimally small volume. The universe is created in, or makes a transition to, an anomalous state in which a vast energy-density resides in the vacuum. This acts like a powerful repulsive force, totally overwhelming gravity. As I explained in Chapter 4, inflation solves what is

known as the 'horizon' problem, that when we look in opposite directions on the sky at the microwave background, we are seeing regions of
the universe that are outside each other's horizon, the limit from which
information travelling at the speed of light can come. In the standard
model of the universe, these regions have never been capable of influ-
encing each other prior to the epoch at which we see them. How then
could they be so similar? Inflation solves this problem because the
whole observable universe has inflated from a very small volume all of
which was easily in communication at early times.

The version of inflation invented by Alan Guth did not quite
work, though. In his model it turned out that the universe would
emerge from the inflationary period as a network of separately inflating
'bubbles' and these would not merge together and evolve to the relatively smooth universe we see today. The solution to this problem,
which was called the 'graceful exit' problem, because it was hard to get
the universe to make a graceful exit from the inflationary phase, was
found in 1982 by the Russian cosmologist Andrei Linde, and independently by the Americans Andreas Albrecht and Paul Steinhardt. The
trick was to have a much more gradual transition from the state of
'false' vacuum. This meant that the inflation continued for longer and
the whole observable universe today would have developed from a single
inflationary bubble. The disadvantage of this version of inflation is that
it involves rather fine tuning of the inflationary force, and there is no
real explanation of why this should be so.

One of the consequences of the inflationary scenario is that we
would expect the universe to have an average density very close to the
critical value, within a small fraction of a percent, in fact. Now, from
the abundances of the light elements, helium, deuterium and lithium,
we can deduce the density of the universe in ordinary, baryonic matter,
because these light elements were made during the fireball phase of the
Big Bang and the amount produced depends sensitively on the density

———————▶

*Fig. 10.3 The optical galaxy distribution: (a) a survey of galaxies in the
southern sky made by George Efstathiou and colleagues from the Palomar
Sky Survey using the Cambridge Automatic Plate-measuring Machine, (b)
the Center for Astrophysics galaxy redshift survey, with the 'Great Wall' of
galaxies (c) John Huchra and Margaret Geller, of the Harvard-
Smithsonian Center for Astrophysics.*

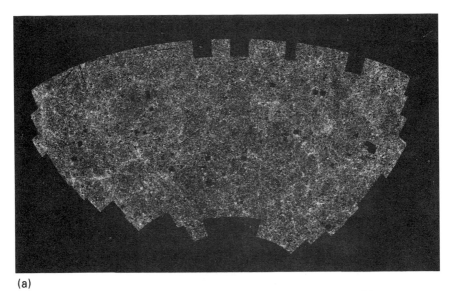

(a)

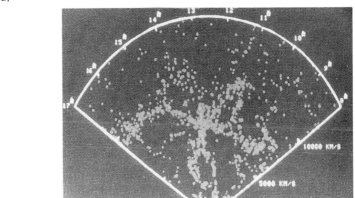

(b)

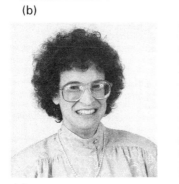

(c)

of the universe. The result of calculations of the cosmological nucleosynthesis, which took place during the first 100 seconds after the Big Bang, is that the mean density of the universe in the form of baryons is only about 3% of the critical value (see p.127). This is not far off the value of about 10% found by adding up all the matter in galaxies, including their dark halos. Thus the dark matter in the halos of galaxies could probably be baryonic, for example Jupiter-sized objects, brown dwarfs (up to 80 Jupiter masses) or massive black holes of around a million times the mass of the sun. But since inflation predicts that the universe should still be very close to the critical density, it requires that 97% of the matter in the universe today should be in some invisible, non-baryonic form.

We have already seen that cosmologists need non-baryonic dark matter to explain how galaxies emerged from the highly smooth state that matter and radiation were in when they decoupled from each other. A universe dominated by dark matter is also a strong prediction of inflationary cosmology. The cold dark matter scenario ties together particle physics, cosmology and the theories of galaxy formation and clustering into an elegant package. It was this that accounted for the prestige the theory enjoyed amongst particle physicists and astronomers alike.

Problems for the cold dark matter scenario

Why is this theory in difficulty today? No matter how impressive a theory is, it has to fit in with what we actually see. Let us look first at some of the observations of the galaxy distribution made over the past decade. In 1983 John Huchra, Marc Davis and their colleagues at the Harvard-Smithsonian Center for Astrophysics measured the distances of over 2000 galaxies by observing their redshifts with groundbased optical telecopes. The light from galaxies is shifted towards the red end of the visible spectrum (to longer wavelengths) because the galaxies are moving away from us. In an expanding universe the recession velocity is proportional to the distance of the galaxy. Huchra and his colleagues were able to map the concentration of galaxies centred on the Virgo cluster, known as the Local Supercluster, and to explore galaxy clustering on scales out to 100 million light years.

More recently, in 1990, deeper surveys by Margaret Geller, also of the Harvard-Smithsonian Center for Astrophysics, John Huchra and their colleagues revealed the existence of the 'Great Wall', a sheet of galaxies 300 million light years long connecting the Coma and Hercules superclusters. They also confirmed the existence of substantial voids almost free of galaxies. The first such void to be identified was the Bootes void, found in a deeper survey of a small area of the sky by Robert Kirschner, also at Harvard, and his colleagues in 1981. These walls and voids seemed unexpected within the framework of a galaxy formation scenario in which galaxies and clusters form by gravitational aggregation from random density fluctuations, as in the cold dark matter model. However, the cold dark matter theory's supporters argued that features of this kind were in fact apparent in their numerical simulations of the galaxy distribution. Whether the observed galaxy distribution and the simulations agreed or not remained a matter of dispute. The discovery of one or two walls and voids was not sufficent to refute the cold dark matter model.

Coup de grace for cold dark matter

The coup de grace for the cold dark matter scenario was delivered by the QDOT redshift survey of IRAS galaxies, which I described in the last chapter. This survey allowed us to make the first deep maps of the galaxy distribution covering virtually the whole sky and these maps did not agree with the predictions of the cold dark matter (CDM) model. Of course not all types of galaxy are strong enough infrared emitters to have been detected by IRAS. In particular, elliptical galaxies, in which very little star formation is going on today, and which comprise about 20% of the galaxy population, are generally very weak infrared emitters. Ellipticals tend to be concentrated towards the cores of dense clusters of galaxies and so IRAS samples these regions poorly.

My research student Will Saunders, as part of his PhD project,

Fig. 10.4 The QDOT all-sky IRAS galaxy survey. The nearer galaxies are shown with larger symbols. White areas were excluded from the survey.

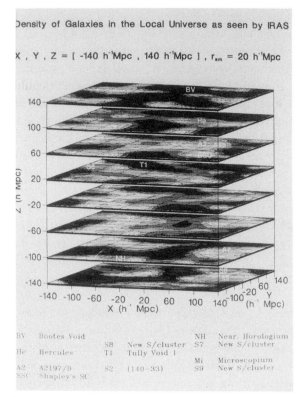

Fig. 10.5 (a) QDOT three-dimensional maps of the galaxy density distribution. A series of horizontal slices are shown. Light areas are regions of high galaxy density (clusters) and dark areas are regions of low galaxy density (voids).

(b) Comparison of QDOT density fluctuations on different scales with cold dark matter model prediction.

(a)

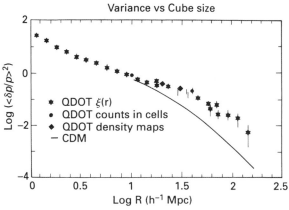

(b)

had derived three-dimensional density maps from the IRAS galaxy distribution. These are really the first reliable, deep density maps of the universe available to astronomers. After completing his PhD and moving to Oxford, Will transformed these maps to vivid colour pictures. He also generated a histogram of the density distribution which could then be compared to the predictions of models, for example the CDM model. George Efstathiou had already devised a new statistical test ('counts in cells') to perform this kind of test and the results, published by George and the QDOT team in the Royal Astronomical Society's journal *Monthly Notices* in December 1990, were clearly inconsistent with CDM.

Will's histograms, and the statistical tests he calculated from them, showed the result in a much more transparent way. The universe is lumpier on the large scale than predicted by the current version of the cold dark matter theory. The paper was submitted to *Nature*, who decided to print the colour density maps on their front cover. They also ran a feature in their 'News and Views' column stating that this paper meant the end of CDM and put out a sensational press release. The result was that reports on the paper appeared on front pages of newspapers round the world, including *The New York Times* and *The Times* of India.

One reason this paper attracted so much attention was that the QDOT team included several of those who had been most active in promoting the CDM theory. Carlos Frenk felt very strongly that the reporting of our *Nature* paper had overstated the problem for the cold dark matter theory and he persuaded George Efstathiou and Simon White to join him in a letter to *Nature* pointing this out.

Some months later, BBC television's *Horizon* programme made a one hour programme centred round our work. The director, Martin Chilcott, had secured the funding for the programme on the basis of the concept 'Big Bang theory in trouble'. When he came to see me about this idea, we went to a local pub for lunch and I spent three hours trying to convince him that the real story was that it was the cold dark matter scenario for galaxy formation, not the Big Bang theory itself, which was in trouble. I gave him a copy of a piece I had written for the magazine *New Scientist* which was to appear shortly explaining all this. Other members of the QDOT team also explained the story to him. When I next saw Martin some months later for the filming of the

THE NEW YORK TIMES (collage) 3/1/91

The New York Times

"All the News That's Fit to Print"

NEW YORK, THURSDAY, JANUARY 3, 1991

50 cents beyond 75 miles from New York

L.CXL.... No. 48,469   Copyright © 1991 The New York Times

Astronomers' New Data Jolt Vital Part of Big Bang Theory

By JOHN NOBLE WILFORD

A critical element of the widely accepted Big Bang theory about the origin and evolution of the universe is being discarded by some of its staunchest advocates, throwing the field of cosmology into turmoil.

According to the Big Bang theory, matter from the explosive moment of cosmic creation originally was evenly spread throughout the universe. But galaxies tend to be clumped together, an awkward fact that astronomers have sought to explain by assuming that cold invisible matter is a major attractive force.

The cold dark matter model, as it is called, accounts well for local clustering but could not explain the giant superstructures recently found in galactic surveys, like the "great wall," a string of galaxies that stretches across the sky for at least a half billion light-years.

A new analysis of a highly accurate survey by the Infrared Astronomical Satellite now shows the universe to be full of such superstructures and companion supervoids. A major problem is that these structures appear to be far too vast to have formed since the Big

formation." In a report being published today in the journal Nature, they said the theory in its present form must be abandoned.

The journal noted that the report by Dr. Will Saunders of Oxford University and colleagues "is all the more remarkable for coming from a group of authors that includes some of the theory's longtime supporters."

'Too Good to Be True'

With its repeated inability to reconcile the evidence and the theory, cosmology is in disarray, trying to patch together a modified version of the theory, testing alternative concepts that had been set aside and looking for entirely new theories to explain how the universe got to be the way it is observed today.

"We're floundering around with lots of ideas," said Dr. Alan Dressler, an astronomer.

Continued on Page A19, Column 1

| News Summary | A2 |
| Editorials/Op-Ed | A20-21 |

New Data Cast Doubt on Part of Big Bang Theory

Continued From Page A1

trophysicist at the Carnegie Institution of Washington, regretting the apparent demise of a theory that was widely held for more than a decade. "The one idea that was too good to be true turned out to be too good to be true."

Dr. Jeremiah P. Ostriker, an astrophysicist at Princeton University, said the results of the survey by the Infrared Astronomical Satellite, showing the wider distribution and greater density of galactic clusters, "sounded the death knell" to the cold dark matter theory.

Validity of Big Bang

Dr. Edmund Bertschinger, an astrophysicist at the Massachusetts Institute of Technology, cautioned "Most cosmologists admit that the cold dark matter theory is in very serious trouble. But not all of them would agree that it is dead. I would say it's premature to say we must reject cold dark matter."

Despite the confused state of their field, cosmologists insist that there is no reason to abandon the overall Big Bang theory or to question the premise that lay at the foundation of the troubled cold dark matter model. Scientists continue to estimate that, judging from gravitational effects on galaxies and other large cosmic structures, 90 percent to 99 percent of the mass in the universe is hidden from view. This is the so-called missing mass or dark matter.

To explain how this matter could be

Although the model was considered successful in accounting for small-scale structure, conglomerations of stars in galaxies, astronomers have begun to find that the texture of the cosmos runs to ever larger structures — clusters of galaxies, superclusters, an extremely long sheet of galaxies dubbed the "great wall" and possibly concentrations of unseen matter so massive that it exerts a gravitational pull on the Milky Way galaxy. Like the great wall, this gravitational force, called the "great attractor," seems too

Old ideas on the universe's birth cannot explain new findings.

large to have formed in the time believed to have elapsed since the Big Bang, from 10 billion to 20 billion years ago.

A Definitive Test

As Dr. Saunders and his colleagues observed, structures on this large scale are too numerous and distributed too evenly to be explained by the cold dark matter theory. The evidence, in their data, obtained six years ago, are considered a definitive test of the theory because they are drawn from an all-sky survey.

Dr. Bertschinger said one approach may be to modify the existing theory, changing some of the assumptions about cosmic densities and variations in density fluctuations that could have caused matter to congregate on such large scales.

"But then the whole model starts looking very baroque," he said. "Most scientists would prefer to think that God had a simpler design for things."

One widely held refinement of the Big Bang — the inflation theory proposed by Dr. Alan Guth of M.I.T., which assumes that the universe underwent a period of phenomenal growth in the first fraction of a second after the primeval explosion, then settled down to the present expansion rate — is just such as modification. The inflationary growth established energy densities and other preconditions for the rapid evolution of galaxies and larger structures.

Although most cosmologists are confident that the Big Bang took place, Dr. Bertschinger said they are questioning the inflation theory.

Working with engineers, Dr. Ostriker has developed what he says is the most advanced computer model so far for testing various theories for the formation of structure in the universe. Those theories can be put into the computer to see what kind of universe they lead to, and whether that universe is like the one that exists now.

Back to 'Hot Dark Matter'

One idea that is likely to be revived, Dr. Ostriker said, is the hot dark matter.

EVENING STANDARD 4/1/91

Space survey gives world astronomers a cosmic jolt

by Mark Long

ASTRONOMERS were today reeling from new evidence casting doubt on all previous knowledge about the formation of gal-

---

THE DAILY TELEGRAPH   4/1/91

THE DAILY TELEGRAPH, FRIDAY, JANUARY 4, 1991   17

All that twinkles . . . is just like grandma's gravy

Astronomers are agog at lumps in the universe, as ERIC BAILEY finds

SO THE universe is lumpy. It is full of gobbets of gal-   sional but the team still needed to plot exactly   away

---

Bang goes a theory of the universe

We live in an unexpectedly lumpy universe. Everyday objects are lumps of matter, and astronomers have known for centuries that outside the earth, matter joins together on an increasingly large scale to which

dark matter model can be bent to take account of the new observations, because it doesn't have any free parameters to tweak. But I think someone will come up with a brilliant lateral leap of logic as

FINANCIAL TIMES   4/1/91

"WILL YOU TRY TO BE ROMANTIC AND STOP CALLING IT LUMPY?"

QMW scenes, the script followed my article reasonably accurately. The filming itself was quite an ordeal, taking over 12 hours for a total of about 20 minutes on the screen. Martin wanted to make a film that would look good, as well as explaining the science, so it took at least an hour to prepare the lighting for each shot. By using strong lights shining into the window of my office to simulate the California sunshine, and by covering up the obviously British electric sockets and telephone, my office was transformed into the office I had used at JPL in 1983. I was filmed driving up to the front of the college and parking (though in fact this is a strictly no-parking zone), then disappearing through a door which is in fact always kept locked (I had to walk up to the door and then hide against the doorpost). When filming ended late that evening, I returned exhausted to my car, which I had reparked in a back street, to find that it had been stolen.

When the programme was shown, I was at first disappointed to find that in the final editing and commentary (narrated very well by Tim Piggott-Smith), the 'Big Bang in trouble' slant had resurfaced. The stormy week in La Palma which Andy Lawrence and I had spent in January 1987, and which I described in the previous chapter, had become a little bit of light rain which soon cleared. And as for the observing to loud rock music, one of the most popular moments in the programme with many who saw it, this was pure fiction (at least when I was observing). However, science journalists and science producers always have great difficulty getting space for articles or programmes about science, so I guess scientists have to put up with some exaggeration and inaccuracy if they want to see science reported.

Were our maps of the density of galaxies really the death of cold dark matter? Our analysis of the pull acting on our galaxy using IRAS galaxies had shown that the universe has a density close to the critical value and is therefore 90% or more composed of some form of dark matter. We knew we needed this dark matter to account for the formation of galaxies despite the astonishingly smooth microwave background radiation. The issue is therefore not *whether* there is dark matter in the universe, but what form it takes. CDM is very effective at explaining the formation of galaxies and clusters. But it seems to have a problem explaining the lumpiness we see on large scales.

◄————————

Fig. 10.6 Headlines on the 'death' of the cold dark matter model.

There were three choices. The first was to modify the CDM scenario to give more structure on large scales. For example several people, including Carlos Frenk, have explored the idea that the mechanism for 'biasing' galaxy formation (see p.148) could be adjusted to improve the fit of the CDM scenario to the data. Another modification to the CDM picture, proposed by George Efstathiou and colleagues, is to suppose that there might still be a significant cosmological repulsion in the universe today. This increases the time-scale for gravity to act and build up significant large-scale structure. The second type of solution was to postulate some new dark matter ingredient in the universe, for example a component of hot dark matter, to account for the large scale structure. This was the idea that I personally favoured and was to pursue even more strongly when the COBE results were announced just over a year later. Finally, the third choice was to try to come up with some completely new framework for galaxy formation and clustering. Theorists are still vigorously pursuing all three avenues.

There were two sad ironies about the day that our *Nature* paper was published. In the morning, I attended a meeting of the UK's Space Science Programme Board, at which we learned of the severe scale of the cuts that the Government and the Science and Engineering Research Council (SERC) had decided to inflict on ground-based and space astronomy. So on the day that we demonstrated rather well the effectiveness of Britain's role in both areas of astronomy, the possibility of such a project being pursued in the next decade severely diminished. And then in the afternoon, Andy and I drove up to the funeral of Michael Penston, who had been in with us at the beginning of the project. His premature death from cancer deprived us of a friend and of one of astronomy's brightest and wittiest stars. He was a strong believer in the popularization of astronomy and would have enjoyed the wide coverage.

From quasars to ultraluminous infrared galaxies

THE REDSHIFT SURVEYS of IRAS galaxies pursued by my group at QMW and our collaborators was to result in yet another spectacular discovery, the most luminous galaxy known in the universe, which may be a galaxy in the process of formation. I mentioned in Chapter 7 the discovery of 'ultraluminous' IRAS galaxies, with infrared luminosities one hundred times greater than the entire luminous output of our galaxy. This new galaxy, with the unprepossessing name of IRAS F10214+4724, was a further one hundred times more luminous than these. To set the scene we need to look at the story of the objects which had hitherto held the record for most luminous objects in the universe, the *quasars*.

The discovery of quasars

On August 5th, 1962, three Australian radio-astronomers watched the moon eclipse a source selected from the 3rd Cambridge Catalogue of radio sources. By timing the instant of disappearance and reappearance, they hoped to establish the position of a radio source for the first time to an accuracy of one second of arc ($\frac{1}{3600}$th of a degree). But the observations turned out to be far more significant, for the source Cyril

159

Hazard and his collaborators had selected was 3C273, the visually brightest member of a class of objects which has become known as quasars, and which include the most powerful and distant sources known in the universe.

The accurate position allowed Cyril Hazard and his colleagues to identify the radio source as an unusual 13th-magnitude star that has a jet of light sticking out from the side of the stellar image. This was not the first time that radio sources had been found to be associated with stars. Allan Sandage and Tom Matthews had in 1960 identified a number of radio sources with stars, but had not been able to make sense of their spectra. At the meeting of the American Astronomical Society in New York in December that year, Allan Sandage reported that the radio source 3C48 was identified with 'a sixteenth magnitude object in Triangulum that appears to be the first case where strong radio emission originates from an optically observed star'. The optical spectrum showed 'a combination of absorption and emission lines unlike that of any other star known. . . There is a remote possibility that it may be a very distant galaxy of stars: but there is a general agreement among the astronomers concerned that it is a relatively nearby star with most peculiar properties.'

Fig. 11.1
The first quasar,
3C273.

Cyril Hazard suggested to Maarten Schmidt of the California Institute of Technology that it would be worth taking a spectrum of the star identified with 3C273 with the 200 inch telescope. The spectrum Maarten Schmidt measured was dominated by four strong emission lines and it looked quite different from that of any star, nebula or galaxy known. Stars show characteristic spectra with emission or absorption lines depending on the temperature of the star. The spectra of gaseous nebulae like the Orion Nebula show strong emission lines at wavelengths characteristic of the elements in the gas (lines of hydrogen and oxygen are especially prominent in the visible band). And galaxies tend to give a rather featureless spectrum with weak stellar absorption lines of calcium and sodium. Although the observed wavelengths of the lines in the spectrum of 3C273 did not correspond with any known lines, Maarten Schmidt recognized that they corresponded to spectral lines from the Balmer series for hydrogen, but shifted towards the red by 15.8%. If the redshift were due to the expansion of the universe, then the object was very distant and a hundred times more luminous than the most luminous galaxy known. Following Schmidt's lead, Jesse Greenstein and Tom Matthews, also from Caltech, were immediately able to identify the unusual emission lines in the spectrum of the stellar object which in 1960 had been identified the radio source, 3C48. In this case the lines were due to ions of magnesium, neon and oxygen redshifted by 36.8%. The drama of the quasars had begun.

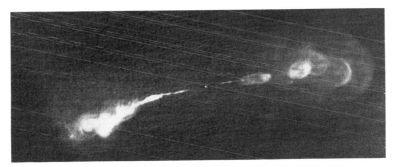

Fig. 11.2 Example of a double radio-source, the radio-galaxy 3C348.

The papers by Hazard and his co-workers, by Schmidt, and by Greenstein and Matthews, were published in *Nature* on March 16th, 1963. In retrospect it seems surprising that the papers of Hazard and Schmidt were not combined, for Hazard and his colleagues deserved to share some of the credit for the discovery of quasars with Schmidt. Today the widespread availability of time on large telescopes has changed the scientific mores. Hazard would have asked the current observer on a 4-metre telescope to take a spectrum and that observer would probably have sent the data to Hazard without necessarily expecting to have his or her name on the discovery paper.

By December 1963, when the first Texas Symposium on Relativistic Astrophysics was held in Dallas, the name 'quasistellar radio source' had been shortened to QSR, or quasar, and astronomers had identified many of the main characteristics of this new class of object. The spectra of quasars typically showed very broad emission lines, characteristic of hot gas in relative motion at speeds of tens of thousands of kilometres per second. These lines tended to be redshifted by large amounts. Quasars with spectral lines shifted by over 200% in wavelength, so the observed wavelength was over three times that emitted by the quasar, were soon to be found. In the spectrum of these quasars, for example, the far ultraviolet Lyman-alpha line of hydrogen emitted at a wavelength of 1216Å began to be observed at the blue end of the visible spectrum, which begins at a wavelength around 3600Å. Quasars tended to emit very much more radiation at ultraviolet and infrared wavelengths compared with normal galaxies. They showed optical variability on a time scale of years and, possibly, days, implying that the objects were not much larger than the solar system; and a compact stellar appearance at visual wavelengths, occasionally accompanied by jets or wisps of nebulosity. It was soon discovered by Allan Sandage that most quasars (over 80% of them) were not in fact strong emitters of radio waves.

It took some years to establish for certain that the cause of the high redshifts of quasars is simply the expansion of the universe. As I described in Chapter 4, enough examples were eventually found of quasars in groups and clusters of galaxies with the same redshift to establish the cosmological nature of their redshifts beyond reasonable doubt. It also began to be realized that quasars are some kind of energetic event or outburst in the central nuclei of galaxies. They began to be seen as part of a wider class of active galactic nuclei, or AGN, which included less dramatic phenomena such as radio-galaxies and Seyfert galaxies. In the 1970s, as a result of a series of X-ray astronomy space missions, starting with the Uhuru satellite in 1970, it was discovered that quasars are strong sources of X-rays. Improved maps of the radio emission from those quasars which are strong radio emitters showed that many have a double-lobed structure clearly powered by a narrow, double beam emerging from a central point.

When quasars were first discovered they were thought by many to be due to explosions in galactic nuclei, but this idea gradually came to be replaced by the idea that quasars are powered by gas falling into massive black holes. The possibility that black holes might power quasars was first suggested by the Russians Yakov Zeldovich and Ivor Novikov, and by the American Ed Salpeter as early as 1964, and the idea was developed further by Donald Lynden-Bell in 1969. The X-ray emission and beamed radio structure were important factors in establishing this picture. Martin Rees and Roger Blandford, at Cambridge, were the first to show in detail how a black hole into which gas is falling from outside could give rise to a double-beamed radio source. In some quasars, for example 3C273 itself, the beam is believed to be oriented close to our line of sight, for over a period of years clumps of emission are seen moving out along the beam at speeds which appear to

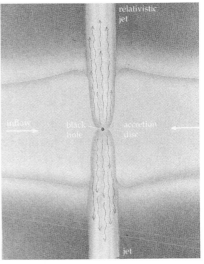

Fig. 11.3 Model for central region of a quasar, showing a massive black hole surrounded by an accretion disc, which radiates at X-ray and ultraviolet wavelengths. Twin, oppositely-directed, relativistic beams emerge along the axis of the disc and create the characteristic double radio-source.

be faster than the speed of light. This is probably an optical illusion caused by the clumps moving at almost the speed of light close to our line of sight (this is a well-known phenomenon of Einstein's Special Theory of Relativity). However, Roberto Terlevich, of the Royal Greenwich Observatory, likes to argue that most of the properties of quasars can be explained by a burst of massive star formation in a galaxy's nucleus, with lots of supernovae explosions going off.

Evolution of the quasar population

My own fascination with quasars began as a research student at

London University in 1964. One of the first assignments suggested by my supervisor, Bill McCrea (now Sir William McCrea), was to test whether the twenty or so quasars then known lay on a great circle on the sky. If this had been the case, this might have been strong support for the idea that quasars are relatively local objects. It proved not to be so, and a year or so later Michael Penston and I showed that quasars are distributed fairly uniformly across the sky. However, I became interested in the question of how quasars are distributed with depth in space. Are they uniformly distributed or do we, as we look further out into the universe and therefore further back in time, see evolutionary changes in the quasar population?

Malcolm Longair, a member of Martin Ryle's radio-astronomy group at Cambridge, showed in 1966 that the radio-source population was undergoing significant evolution, changing its properties with time. He believed that this was due to evolution in only the most luminous radio-sources, which he identified as the quasar population, whereas I thought the evolution also affected the more numerous and less luminous radio-galaxy population.

If we think of a quasar outburst as a brief phase, perhaps repeated intermittently, in the life of a galaxy, then we may see evolution for several different reasons. In the past a greater fraction of galaxies may have gone through a quasar phase than do so today; or the phase may have lasted longer. In both these cases we would expect to see an increase in the number of quasars per unit volume (after correction for the expansion of the universe) as we look out to great distances, and hence back to much earlier times. Alternatively, the luminosity of the quasar phase may have been greater in the past. The direct evidence from the quasars themselves allowed many different types of evolution. But when I took into consideration the interpretation of counts of faint radio-sources I concluded in 1970 that the simplest picture was one in which the typical luminosity of quasars (and luminous radio-galaxies) had steadily decreased with time over the thousands of millions of years of cosmological time over which we are seeing them. Subsequent studies have proved this to be the case and recently it has become clear that the radio-quiet quasars show the same evolution. The discovery of evolution in the quasar population in the late 1960s came at a time when there was still vigorous debate between supporters of the rival Big Bang and Steady State theories and it was one more nail in the coffin of the Steady State theory.

The discovery of ultraluminous IRAS galaxies

What does this evolution mean? Why should the typical luminosity of quasars have been greater in the past? Despite decades of work on quasars there has been no real progress in finding the physical cause of this evolution. To find the answer to this question we have to return to the story of the galaxies detected by IRAS, which I started to describe in Chapter 7. Although many of the first galaxies which we identified from the IRAS survey were normal spiral galaxies, a few were more unusual. One of the most unusual turned out to be number 220 in Halton Arp's Atlas of Peculiar and Interacting galaxies. The total infrared luminosity of this galaxy turned out to be more than a million million times that of the sun, or nearly a hundred times the total output of our own Galaxy. Such a luminosity placed this galaxy in the same class of power output as the quasars. The key difference though is that whereas the quasars put out most of their power at X-ray and ultraviolet wavelengths, the power of Arp 220 is almost entirely radiated in the far infrared. Many examples of such 'ultraluminous' infrared galaxies turned up in our redshift surveys of IRAS galaxies. The groups of Bob Joseph, then at Imperial College, London, and Tom Soifer at Caltech, Pasadena, advanced the view that all luminous IRAS galaxies might be due to interactions or mergers between pairs of galaxies. My own group at Queen Mary and Westfield College was doubtful of this theory at first. Within our large body of survey data on IRAS galaxies, we had found only 10–15% of galaxies to be interacting or merging, a figure not very different from that for optical catalogues of galaxies. It turned out, though, that when attention was focused on the real monsters, the Arp 220 types, the proportion of these that were interacting, merging or peculiar, was indeed very high.

We also found ourselves disagreeing with Caltech on another issue, the fraction of luminous IRAS galaxies that had 'active' quasar-like nuclei. Again we found only 20% or so, compared to 100% found by Caltech. To some extent this is because we were studying more distant galaxies, whereas Caltech had selected a much nearer sample for study, so weak activity can be found in these more easily. But these studies raise the very interesting question: what is the relationship between ultraluminous infrared galaxies and quasars? My view has

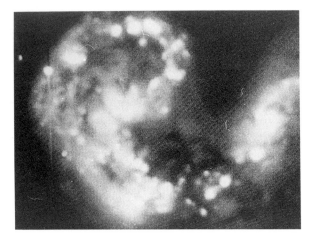

Fig. 11.4
(a) Examples of
interacting galaxies
detected by IRAS,

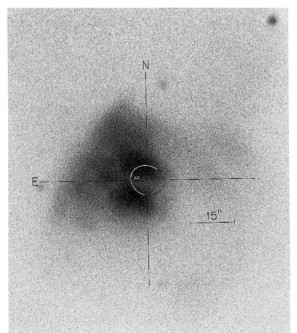

(b) Arp 220.

always been that almost all the emission we see at the far infrared wavelengths of 50–100 microns in luminous IRAS galaxies is due to tremendous bursts of star formation, or 'starbursts', in the nuclei of these galaxies. The Caltech study raises a rival point of view, that these may be quasars immersed in a thick cloud of dust. So far this issue is not resolved, though recent radio studies by Jim Condon of the US National Radio Astronomy Observatory at Charlottesville and his colleagues appear to support the starburst interpretation for the majority of ultraluminous infrared galaxies.

These studies have to be seen against the background of the continuing developments in work on quasars and radio-galaxies. At one time it was thought that very few quasars remained to be found at redshifts greater than 2 (200% shift in wavelength towards longer wavelengths), but careful and systematic studies throughout the 1980s managed to keep on discovering higher redshift quasars. Especially notable was the success of the group of Richard McMahon and Mike Irwin from Cambridge in discovering high redshift quasars with the relatively small 2.5 metre Isaac Newton telescope on La Palma. Today the highest redshifts known are close to 5 and there is no reason to suppose that the limit has been reached yet.

In the same period very careful identification of the few remaining unidentified sources in the famous 3rd Cambridge Catalogue of Radio Sources yielded several radio-galaxies with redshifts in the range 1-3 and beyond. Could these galaxies be the fabled protogalaxies, galaxies in the process of formation, which astronomers had searched for for so long? Unfortunately studies of the optical spectrum seem to show that the stars in these galaxies, while young compared to the sun, are still a billion years old or more, so the galaxies can not really be called protogalaxies.

IRAS F10214+4724, most luminous galaxy in the universe

On June 27th 1991, I published a paper in *Nature* with 12 collaborators on the discovery of a new IRAS galaxy, IRAS F10214+4724, which may be the 'missing link' between ultraluminous IRAS galaxies and quasars, and may also be the best candidate yet for a galaxy close to the moment of formation. The story of this discovery illustrates several aspects of modern science: the large collaborations and powerful resources required, the element of painstaking and systematic work, and the ingredient of pure luck, which nevertheless has to be seized.

It began with the approaching completion of the QDOT survey in 1988, when I began to look for a new programme. Following the publication of the IRAS Point Source Catalog back in 1984, the US Infrared

Processing and Analysis Center (IPAC), based at Caltech in Pasadena, had began to work on a new project. They decided to commit a major effort to the preparation of a new, much deeper catalogue of infrared sources. The idea was to combine all the data from the IRAS survey into a map of the sky and search for the faintest sources visible in the map. This became the IRAS Faint Source Survey and by 1988 a preliminary version of this new catalogue was available. I was invited to help check out this catalogue and I arranged to visit Pasadena for two weeks in September 1988. I was very interested in doing this because I had been sceptical about the project when it was proposed. I had thought that it would not be possible to detect reliably sources much fainter than those in the Point Source Catalog and I was concerned that the technique of adding all the data together would result in the generation of many spurious sources. On the other hand, if the IPAC team had succeeded, I was keen to use the new catalogue at the earliest opportunity.

As usual on these trips to IPAC, I stayed as close as possible to UK time, coming into work at 4 am. This allowed me to use the computers when there was little demand on them. I studied the source-counts and colours of the sources in the catalogue in areas over the whole sky and studied how these changed with different indicators of emission from interstellar dust. It was clear that in the 60 micron wavelength band, the catalogue was reasonably complete for sources at least a factor of two weaker than the limit of the Point Source Catalog. Although there were many spurious sources included in regions of strong emission from interstellar dust, these were fairly easy to recognize from their infrared colour and other indicators. When I took samples of several hundred sources from different areas of the sky and looked for their counterparts on the Palomar Sky Survey, I found that a high proportion, over 90% at high Galactic latitudes, could be identified with galaxies. The IPAC team had succeeded in generating a powerful new catalogue which would be valuable for cosmological studies. I had to admit to Tom Chester, the leader of the IPAC team, that my scepticism about the project had been unfounded.

I was in contact with some of my UK collaborators by electronic mail during this visit and I started floating the idea of a new galaxy redshift survey based on the Faint Source Survey. The idea was that we would select the cleanest and deepest areas of the sky in the northern hemisphere and study the galaxies in these areas. I suggested that some of the IPAC team might join this collaboration and Carol Lonsdale, Perry Hacking and Tim Conrow agreed to join us. When I got back to the UK we submitted an ambitious proposal to survey 700 square degrees of the sky using five weeks of time on the Isaac Newton Telescope on La Palma to measure redshifts of the bulk of the galaxies, followed by six nights with the 4.2 metre William Herschel Telescope, also on La Palma, to finish off the very faintest of the IRAS galaxies.

Somewhat embarrassingly, this proposal arrived at the allocation committee for the the UK's La Palma telescopes just as I was taking over as chair of the committee. Naturally when you are on such a committee you do not have anything to do with your own proposals and you have to leave the room when they are discussed. Luckily our proposal got the time it needed.

The team of collaborators on our IRAS Faint Source Survey galaxy redshift programme included my IRAS group at Queen Mary and Westfield College and astronomers from Cambridge, Caltech, Oxford and Durham, so the acronym this time was QCCOD. Different teams took part in the five one-week runs with the Isaac Newton Telescope, but the lion's share of the work was done by my postdoctoral assistant Tom Broadhurst and my research student Seb Oliver. After four weeks of measuring redshifts in the springs of 1989 and 1990, we spent a week taking images of the fields of sources we had failed to identify from the Palomar Sky Survey.

The final run with the William Herschel Telescope (WHT) took place in May 1990, with the aim of taking spectra of the very faintest galaxies not accessible with the smaller Isaac Newton Telescope. Tom Broadhurst and I went on this run and we had a list of some 150 sources which still did not have redshifts out of a total sample of 1400 faint IRAS sources. It is always immensely exciting to start work with one of the world's great telescopes, for the possibilty of some startling discovery is always there. We knew that we were looking for the first time at the most distant and faintest of the galaxies detected by IRAS, so with every spectrum we hope to see something unusual. Most, though, continued to be the same kind of starburst galaxies we had seen thousands of times before.

Observing at La Roque de los Muchachos, La Palma

I arrived at the summit towards dinner time after the usual ghastly taxi drive up the mountain. I always ask the driver, Lionel, to go slowly, but I still always seem to end up feeling travel-sick after the endless hairpin bends. The summit looks, as always, breathtakingly beautiful, barren, rugged, with the white domes enclosing the telescopes perched on all the vantage points catching the evening sun, the sea precipitously below us in every direction, the immense silence.

Pieter Morpurgo, Patrick Moore and BBC Television's 'Sky at Night' team are there at dinner. They are making a programme about the William Herschel Telescope and they want to interview me the following afternoon in the control room of the WHT. Michael Penston and I, the two old-timers present, reminisce during dinner about astronomers we have known. That night Tom and I go up with the WHT observers and spend much of the night with them, finding out how the telescope is working and getting used to the night shift. I usually sleep at most four or five hours when I am observing, then wake with my mind full of plans for the next night's work. Most of the next afternoon is spent preparing for our first night's observing. The interview takes only half an hour or so. The first night is in fact lost to low cloud sitting on the mountain. The Royal Greenwich Observatory's photographer is also on the mountain to take some publicity photographs of the WHT and he spends the afternoon of our second night taking photographs of the telescope and the control room, including some of Tom and me pretending to observe. He is anxious to get a

Fig. 11.5 Tom Broadhurst(right) and the author(left) at the William Herschel Telescope the night before the discovery of IRAS F10214+4724.

shot of the telescope against the sunset sky and we begin to get agitated as the prospect looms of losing a few minutes observing time in the twilight. Eventually we chase him off and get started. We have a successful night's observing and get spectra of forty of our sources.

I had invited Patrick Moore and Pieter Morpurgo to join us for the next night's observing. It was meant to be a social invitation but they interpreted it as an invitation to film. There is a certain amount of negotiation about light levels, camera locations, etc., so that they will not interfere in any way with the observing. The Spanish telescope operator is alarmed at the prospect of his friends seeing him on television (he is a graduate student and regards the work as rather beneath him) and he refuses to appear in the shot, so we have to operate the telescope ourselves while they are filming. In the event we have only ten minutes of observation before low cloud descends again on the mountain and we have to stop. The next night is also lost to bad weather, but our last two nights are excellent and we are able to more or less complete our programme.

On the very last night of our forty-night programme, although the weather was fine, we had problems with the computer. The disc where our data was being stored filled up and we needed help to sort this out. On restarting, the telescope computer kept crashing and we had a tense night's observing. Eventually we came to the IRAS source F10214+4724. The uninspiring name is in fact the position of the source on the sky, 10 hours and 21.4 minutes of Right Ascension (longitude on the sky) and +47° 24' of declination (latitude on the sky), with F denoting a source from the IRAS Faint Source Catalog. On the photograph of this part of the sky taken by the Palomar Sky Survey, there are three or four possible optical objects which might be the infrared source. We had already tried two of them during our second night but found them to be foreground stars, almost certainly nothing to do with the infrared emission. We therefore tried a third optical object. Tom was examining each spectrum after the integration had been completed and reporting whether we had detected emission lines and therefore had measured a redshift. If we had been unsuccessful I would see whether there were any further possible objects on our 'finding charts' (these are in fact polaroid snapshots taken from the Palomar Sky Survey prints). In this case Tom called out there was a weird-looking set of emission lines. This almost certainly meant that it was a galaxy with high redshift, since normally we always see the same lines, usually dominated by what is called the Balmer H-alpha line of hydrogen shifted by between 0 and 30% in wavelength. Normally we would have gone back to such an object and taken a further spectrum for confirmation. Unfortunately, the computer crashed once again shortly after this and by the time things were working again, the galaxy had set. We had further excitement later in the night when a diffuse object we had taken a spectrum of slightly earlier seemed to have disappeared. Tom recalled that he had had to keep adjusting the position of the spectrograph slit, as if the object was moving on the sky as we were observing it. It suddenly dawned on us that we had probably been observing a comet which happened to fall at an IRAS position as we oberved it. Regrettably, we could not find the object again as it had moved too far. The spectrum we took showed no emission lines either, so we will never know the truth about this phantom.

Months of detective work

When we got back to London, Tom started to try to analyse the data we had taken, but there were various problems in getting some new software working. High on my list of priorities was to extract the 'comet' spectrum and that of the object with weird spectral lines. It turned out to be some months before we were able to examine this spectrum again. In the end, Andy Lawrence and Tom Broadhurst managed to recognize the pattern of the spectral lines by comparison with spectra of high redshift radio-galaxies. The redshift was a staggering 2.286, compared with the previous highest for a galaxy discovered by IRAS of 0.4. But could we be sure that the object whose spectrum we had measured was connected with the infrared emission? Altogether there were four objects visible on the Sky Survey around the IRAS position, which we had labelled A–D. A and B had turned out to be stars. C was the object we had been targeting when we found the high redshift spectrum. On the WHT guider television it had appeared to be a galaxy, but it was not particularly close to the IRAS measured position, so the probability of its being the correct identification was not high.

One of the members of the QCCOD team, Richard McMahon from Cambridge, was visiting IPAC in Pasadena at the time and he and I were in touch by electronic mail. He suggested that he ask Jim

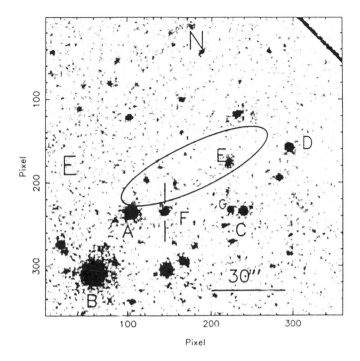

Fig. 11.6 An image of IRAS F10214+4724 made at the Mt Palomar 200" telescope with a Charge Couple Device by Pat McCarthy.

Condon at the US National Radio Astronomy Observatory to make a map of our source with the Very Large Array (VLA) in Soccorro, New Mexico. This is a Y-shaped array of 15 radio-telescopes spread out over 25 kilometres, which is both extremely sensitive and also provides very accurate radio positions. Most spiral galaxies detected by IRAS can also be detected at radio wavelengths with a very sensitive radio telescope like the VLA. If there was a radio source at position C this would help to support the identification. While he was over at the Caltech Astronomy Department, Richard also happened to run into a friend of his, Charles Lawrence, who was on his way up to the 200 inch telescope on Mount Palomar to work with Pat McCarthy. Richard asked for an image to be made of IRAS F10214+4724 with the 200 inch telescope on Mount Palomar, so that we could see if any fainter objects lay close to the IRAS position. This would take them only five minutes or so.

Fig. 11.7 The Very Large Array radio telescope in New Mexico, with which the identification of IRAS F10214+4724 with a redshift 2.3 galaxy was confirmed.

Two days later Richard drove over to the Mount Wilson and Las Campanas Observatories offices in Santa Barbara Street to collect the results from Pat McCarthy. These were perplexing. The optical image from Palomar showed two further very faint galaxies, E and F close to the IRAS position. And the radio map, faxed to us by Jim Condon, showed only one source, located exactly at the position of galaxy F. The evidence was therefore that object C, which we still thought was the high redshift object, was not the correct identification of our infrared source, for otherwise we should have found a radio source at this position.

Meanwhile, Tom Broadhurst had noticed that we did in fact have two spectra of object C, because it happened to lie on the slit when we were observing object A on the second night of our run. He and Seb Oliver had been working on combining the two spectra of the high red-shift galaxy to improve the quality of the spectrum. One Monday morning in December 1990, I came in to find a computer plot on my desk which Tom had left before going off to a conference in Brighton. It was a scan of the total energy falling on the slit of the spectrograph

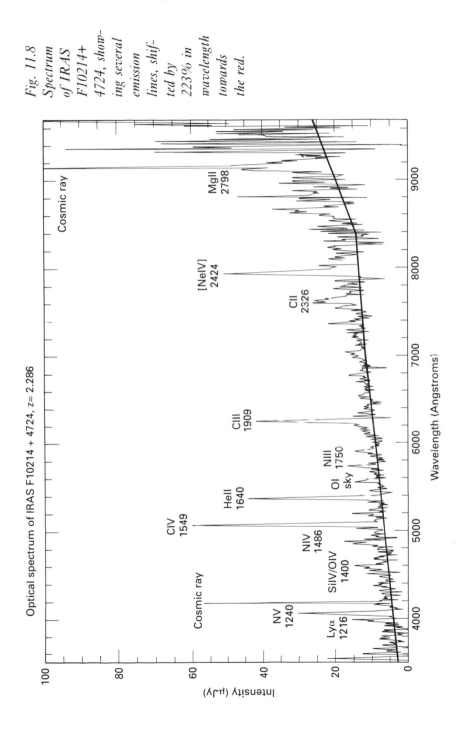

Fig. 11.8 Spectrum of IRAS F10214+4724, showing several emission lines, shifted by 223% in wavelength towards the red.

Optical spectrum of IRAS F10214 + 4724, z= 2.286

as a function of distance along the slit. The brightest peak was star A on the left of the plot. In the middle was a much weaker object, which Tom had labelled 'high redshift object'. And to the right of the scan was a third, brighter source. Pasted on the plot was a message from Tom which read 'What is going on here, which object is this?'.

I counted the number of picture-elements, or 'pixels', between star A and the high redshift object. I then rang up Richard Ellis to check the number of arcseconds per picture-element on this spectrograph (the Faint Object Spectrograph, which had been built by Richard Ellis's group at Durham, in collaboration with Royal Greenwich Observatory). To my amazement and delight, the distance corresponded exactly to the location of the very faint galaxy F. All these months that we had thought we were studying object C, we were in fact studying object F. We had been doubly lucky. Twice, when observing object A and object C, the mysterious object F which was invisible on the Sky Survey and invisible on the William Herschel Telescope's guider TV, had fallen on the slit of the spectrograph. Moreover, at the telescope Tom had selected for examination the spectrum of F, the more centrally located candidate spectrum, rather than of C, the target. The VLA map had proved that this was indeed the identification of the IRAS source. Se we really did have a redshift 2.3 IRAS galaxy. I sent out an electronic mail message to the QCCOD team and, a little later, a first draft of a paper announcing the discovery.

Soon afterwards, I had a request from Tom Soifer, Chief Scientist at IPAC, who had heard about the discovery and was very excited about it, to be allowed to announce it to a NASA board which was deciding the future funding of IPAC. He assured me that the discovery of a high redshift galaxy by IRAS could be critical to the future of IPAC. Given the crucial IPAC role in the discovery, I could not refuse this request, but I was worried that too many people were getting to know about our galaxy before we had published the details. We decided that our paper should be submitted to *Nature* as soon as possible. As I was correcting the proofs of the paper, and preparing a press release to be sent out jointly by Queen Mary and Westfield College and NASA-JPL, I realized that our object was in fact the most luminous object in the universe, exceeding even that of the most luminous known quasars. Although this was not scientifically very important, it would obviously make the discovery more interesting to the public. What was scientifically important was that the far infrared emission was almost certainly emission from dust and implied a huge mass of dust and gas. This alone was enough to make our object look like a massive galaxy in the process of formation. However, the emission line spectrum did look very similar to those of radio-galaxies and Seyferts, so the possibility of a quasar buried behind clouds of dust in the nucleus of the galaxy certainly could not be ruled out. Andy Lawrence and I argued endlessly about this, with him betting heavily on the buried quasar hypothesis, while I strongly backed the protogalaxy idea.

The publication of our paper in *Nature* attracted front page stories around the world as well as coverage on radio and television,

THE TIMES 27/6/91

Dim view of brightest light in the universe

By NIGEL HAWKES
SCIENCE EDITOR

ASTRONOMERS have produced a paradox to stimulate the most jaded appetite: the most luminous object in the universe is too dim to see. The object, IRAS F102214+4724, emits prodigious amounts of energy, around three hundred million million times as strong as the Sun and 30,000 times as powerful as the Milky Way. Yet it is so faint that it was found only by accident when its image crept into a picture being taken of another object.

20 years. Alternatively, it may be a bright quasar embedded in a cloud of dust.

The reason it is so hard to see, in spite of its power, is that it emits its energy in the infra-red spectrum, out of the range of the human eye. Its discovery is reported in today's issue of *Nature* by a team led by Michael Rowan-Robinson of Queen Mary and Westfield College, London. They stumbled across it while looking through a telescope in the Canary Islands for faint sources found by the

83 per cent of the way back to the big bang with which the universe is supposed to have begun. Andy Lawrence, a member of the team, says that they were surprised both by its distance and its brightness. "If it is a proto-galaxy, then it is a very important object indeed," he said. "Nobody has ever found one, and it has become the Holy Grail of astronomy. My view is that it is more likely to be a quasar inside a cloud of dust." Professor Rowan-Robinson, however, believes the galaxy formation

only one, it is hard to come to any conclusions," Dr Lawrence says. "If we had ten or 20 we could compare them and learn a lot about the early stages of the universe."

● When Halley's Comet suddenly increased in brightness last February, there were fears that it had suffered some disaster. David Hughes, of Sheffield university, writing in *Monthly Notices of the Royal Astronomical Society*, has concluded that it must have collided with an asteroid or another comet nucleus.

BRILLIANT! STAR BRITS FIND NEW GALAXY..

By ROGER TODD

BRITISH astronomers have discovered the most brilliant object in the universe

DAILY MIRROR 27/6/91

Britons find galaxy at moment of birth

By Roger Highfield, Science Editor

THE MOST luminous object in the universe, thought to be a galaxy being born, has been discovered by a British-led team of astronomers.

The distant embryonic galaxy is radiating 300 million million times as much energy as the Sun and 30,000 times that of the Milky Way galaxy containing the Earth, it is announced today in Nature.

"This object is exciting and sexy in itself," said Dr Andy Lawrence of Queen Mary and Westfield College, London.

"But if we can find more of this type, we hopefully can pin down how galaxies form and the early history of the universe."

formation or a quasar buried within a dust cloud.

The team, led by Professor Michael Rowan-Robinson of QMWC, which is part of the University of London, includes astronomers from Cambridge, Oxford and Durham Universities as well as from QMWC, and from Caltech, Pasadena, and Charlottesville, Virginia.

They discovered the object with Britain's 4·2 metre William Herschel telescope on La Palma in the Canaries on the last night of a 40-night observing campaign with optical ground-based telescopes, studying very faint sources detected by a satellite.

The object was too faint to

THE DAILY TELEGRAPH 27/6/91

Fig. 11.9 Headlines on discovery of IRAS F12014+4724.

including a send-up of the story on the BBC radio satirical programme 'Week Ending'.

Since our announcement, IRAS·F10214+4724 has been observed using many telescopes around the world. As soon as we realized its significance we had arranged for it to be imaged from the ground at radio wavelengths with the Very Large Array in New Mexico, at submillimetre wavelengths with the James Clerk Maxwell Telescope on Mauna Kea, Hawaii, at near infrared wavelengths with the UK infrared Telescope also on Mauna Kea, and at optical wavelengths with the 200 inch telescope on Mount Palomar. We commissioned observations at X-ray wavelengths with the ROSAT satellite, and also reobserved it with the William Herschel Telescope on La Palma to get an improved optical spectrum. Many other astronomers also observed it with a variety of telescopes. One of the most exciting discoveries was the detection of the molecule carbon monoxide (CO) by Robert Brown and Paul Vanden Bout at the National Radio Astronomy Observatory's Millimeter Wave Telescope on Kitt Peak, Arizona. This took me by surprise as I had not imagined current receiver sensitivities were such that so faint a source could be detected. The detection of carbon monoxide has been confirmed by several other groups working at the French-German IRAM telescope and at the Japanese Noboyama telescope. The total mass of molecular gas implied by the CO detection is vast, more than one hundred billion times the mass of the sun, far greater than has ever been seen in a galaxy before and strongly suggestive of the protogalaxy picture. But other observations (for example, the fact that the optical light is strongly polarized) support the buried quasar idea, so perhaps both pictures are correct. We have continued our search for other such galaxies, and have mapped with the VLA many faint IRAS sources which do not seem to have any galaxy identification on the Palomar Sky Survey to try to locate other examples. However, we have not managed to find another galaxy of this type yet.

COBE discovers the cosmic ripples 12

THE MAIN THEME of this book has been the search for an explanation of how galaxies, clusters of galaxies and even larger structures have been formed in a universe which was initially incredibly smooth and uniform. Current ideas on how galaxies formed, based on the idea of a universe filled with cold dark matter, were thrown into disarray by the discovery of excessive amounts of structure on very large scales in our IRAS galaxy surveys. In this chapter I describe the impact of another tremendously fundamental discovery of the past few years, that of 'ripples' in the microwave background radiation made by the Cosmic Background Explorer satellite (COBE). These were the first sign of the seeds that would ultimately result in the formation of structure in the universe.

The birth of COBE

In 1974 John Mather of the NASA Godard Space Flight Center and six others, including David Wilkinson of Princeton, Ray Weiss of MIT (who was to become the chairman of the COBE Science Team) and Mike Hauser, a fellow IRAS science team member, submitted a proposal to NASA for a space mission to study the spectrum and isotropy

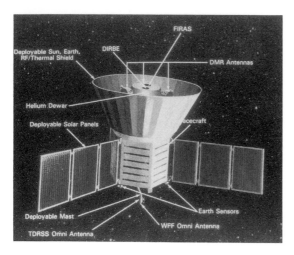

Fig. 12.1 The Cosmic Background Explorer (COBE) satellite.

of the microwave background radiation. Two other rather similar proposals arrived at NASA headquarters at the same time, one from Sam Gulkis and colleagues at JPL, and the other from George Smoot and colleagues at Berkeley. These proposals were merged to form the Cosmic Background Explorer, or COBE. The three main instruments were to be the Far Infrared Absolute Spectrometer (FIRAS), led by John Mather, which was to measure the spectrum of the microwave background radiation, the Diffuse Infrared Background Experiment (DIRBE), led by Mike Hauser, to map the far infrared background radiation, and the Differential Microwave Radioameter (DMR), led by George Smoot, to measure the small scale anisotropy of the microwave background.

Although originally proposed for launch on a Delta rocket, NASA decided to switch the mission to a Shuttle launch. The mission nearly went under with the Challenger disaster of 1986. With considerable modifications and redesigns it was switched back to a Delta rocket and finally launched on November 18th, 1989. An early result from John Mather's FIRAS instrument was a beautiful confirmation of the blackbody nature of the spectrum of the microwave background radiation (see p.121), which showed beyond reasonable doubt that this radiation is the relic of the fireball phase of a Big Bang universe.

The COBE ripples story breaks

The announcement of the discovery of small-scale fluctuations in the cosmic microwave background radiation on Thursday April 23rd, 1992, took most of the world's cosmologists by surprise. Although rumours had abounded that the COBE team might be about to make an announcement, they had managed to keep their discovery secret since the previous September. Almost uniquely in the history of major discoveries, the first news reached scientists via 'the wires' of the press agencies. In this age of electronic mail, scientists chat to each other round the world every day via computer networks, so usually hear scientific news before it is announced to the media. This was the beginning of an exciting few weeks.

That Thursday was an especially testing day, since astronomers had to respond to the journalists off the cuff, without knowing the details of the COBE announcement. What happened in the UK was fairly typical of what was going on all around the world, though for reasons that will become clear, the coverage in the UK was particularly frenzied. At 11 o'clock that morning, Susan Watts, the technology correspondent of *The Independent* newspaper in London, rang me about a story coming in 'over the wires'. COBE is to announce the discovery of

Fig. 12.2 COBE map of the microwave sky showing our Galaxy, cosmic ripples and instrumental noise.

ripples in the background radiation in a news conference at 4 pm. I know nothing about it, of course. She reads the press agency report out to me. It is full of hype – 'the discovery of the century, perhaps of all time', 'the Holy Grail of cosmology', 'English does not have enough superlatives', 'a certain Nobel Prize'. But there was also the crucial piece of information that temperature fluctuations of 30 micro-Kelvins are seen when regions 500 million light years apart are studied. This detail was enough to reconstruct what the COBE team were about to announce and I explained the story to Susan Watts over the next hour or so.

In fact this was a major story and NASA did not need to exaggerate it. Observers have been hunting for these fluctuations since the discovery of the microwave background radiation in 1965. This background gives us a snapshot of the universe 300,000 years after the Big Bang, when the 'fireball' phase ends and matter decouples from radiation. Each time the observers improved the accuracy with which they measured the smoothness of the background radiation, the theoreticians were forced into more exotic explanations of how galaxies could have formed by the action of gravity on such an initially smooth universe. Ten years ago they were forced into the fateful step of assuming that to form galaxies and clusters of galaxies by now, the universe must be dominated by some form of exotic 'dark matter' which would not leave an imprint on the microwave background. Ordinary matter could then accumulate together under the action of the gravity of the lumps of dark matter.

In Chapter 10 I explained that there are two versions of what this dark matter could be. The version which appeared to work best at explaining the distribution of galaxies we see today is 'cold dark matter', composed of hypothetical particles with names like 'axions' or 'photinos', which would be slow-moving (hence 'cold') in the early universe. Alternatively, if the particles were neutrinos with a small non-zero mass, they would be moving close to the speed of light in the early universe, hence 'hot dark matter'. At about the same time as cosmologists realized they needed dark matter, particle physicists introduced a dramatic new concept into the Big Bang picture – *inflation*. They suggested that soon after the Big Bang the universe went through a phase of rapidly accelerating expansion, which inflated the currently observable universe through 70 powers of 10 from much smaller than the size of an atom to the size of a tennis ball. The inflationary scenario also required the universe to be mainly full of dark matter. Before April 23rd, 1992, the microwave background smoothness limits were getting close to destroying this whole picture. The COBE discovery confirms that dark matter is needed to make galaxies, but does not tell us precisely how much dark matter there is. It certainly does not allow us to predict that the universe will end in a Big Crunch, as several reports insisted.

The phone continues to ring all day. At 4 pm Tim Radford of *The Guardian* newspaper rings from Edinburgh. Like most of the British science correspondents, he is there covering the Edinburgh Science Fair and a debate about science and religion between the biologist and science writer Richard Dawkins and Archbishop Hapgood. This debate, although of course inconclusive, has had wide coverage in the British media. The ripples story, in seeming to penetrate to the very birth of the universe, seems like a natural sequel. Tim confirms that *The Guardian* will also have the story on their front page. Sky Television ring and ask me to appear on their Breakfast News programme next day. They ask me to provide some pictures and send a motor-bike to my home later in the evening to pick them up.

Later I hear that most cosmologists in Europe and the United States had the same kind of day. At the Space Sciences Laboratories in the University of California at Berkeley, George Smoot's home base, the newspaper and TV reporters lay siege. As George is in Washington taking part in the Press Conference there, the graduate students are the ones that give the interviews.

Friday April 24th: at 6.45 am the car arrives from Sky. On the way I ask the driver to stop at a newsagent and I buy an *Independent*. It has the most wonderful front page spread on 'How the Universe Began'. The piece is mainly based on my conversation with Susan Watts and I am quoted extensively and reasonably accurately.

The taxi driver has a *Daily Mail* on the back seat and so I check that too. Amazingly they have managed to get pieces by Patrick Moore and Stephen Hawking on the COBE results. The driver asks me what I am going to Sky for. I hold up the *Independent* front page and say: 'For this, how the Universe began'. He says that he thought as much, he'd heard about the story on the radio and guessed that this was why he was picking up a professor. He asks what the story is about and realizing that this is a good opportunity to practice what I am going to say later, I try to explain things. He says that this doesn't sound very much like the account given in Genesis. I soothe him with a remark that nomads in the desert might have had some difficulty understanding all the details.

At 8.15 am I do my bit, preceded by the secretary of the British Tall Persons Club and followed by a bee-keeper. I am impressed with the professionalism of the news presenters, who maintain a serene momentum working at a desk covered with a jumble of bits of paper and newspapers. The pictures come up in some slightly arbitrary order chosen by the producer and I talk around them. It is a bit nerve-wracking saying 'I think we have a picture here of the telescope with which the microwave background radiation was discovered' and wondering whether it will actually appear on the monitor.

The taxi drives me across London to the College by a very scenic

THE INDEPENDENT

No 1,722

FRIDAY 24 APRIL 1992

★★★ Published in London   40p

A Nasa spacecraft has detected echoes of the galaxies' birth fourteen thousand million years ago. The discovery about the formation of the stars after the Big Bang has been hailed by excited scientists as the Holy Grail of cosmology. **Susan Watts** and **Tom Wilkie** report

How the universe began

BACK TO CREATION

How the universe evolved from the Big Bang through the first three minutes, to the first clusters of matter 300,000 years on. By 15 billion years humanity had emerged from the dust of the stars

FOURTEEN thousand million years ago the universe hiccuped. Yesterday, American scientists announced that they have heard the echo.

A Nasa spacecraft has detected ripples at the edge of the Cosmos which are the fossilised imprint of the birth of stars and galaxies around us today.

15 billion years
DNA, the molecule of inheritance, and life on Earth emerge

1 billion years

300,000 years
Epoch of recombination: the first ripples of cosmic structure
Discovery announced yesterday

the lumps (stars, planets and galaxies) got into the porridge.

"What we have found is evidence for the birth of the universe," said Dr George Smoot, an astrophysicist at the University of California, Berkeley, and the leader of the Cobe

3 minutes

1 second
Stable subnuclear particles, neutrons and protons, are formed

-270 degrees Centigrade

-255 degrees
Heavy chemical elements produced in gravitational collapse of stars

6,000 degrees

10^9 degrees
Formation of helium and lithium nuclei

10^{10} degrees

leased from the foggy soup of radiation, was set free to be picked up by modern astronomers with their telescopes.

"Further analysis of Cobe's results will shed light on the identity of the mysterious dark matter that we have

these fluctuations should be. How big they are depends on how fast they are able to grow. These results match just the size that the theory predicts. People have been looking for this kind of variation since the 1950s.

However, Professor Arnold Wolfendale, the Astronomer Royal, sounded a note of caution. He said the sci-

GRAPHIC: MICHAEL ROSCOE

route (Hyde Park, the Mall, the Embankment, St Paul's). Susan Watts rings and I congratulate her on the story. She is doing a follow-up piece on the significance of the COBE ripples and what happens next. She says a companion piece is being written on their religious significance. My relief at not being part of that piece is short-lived. *The Daily Telegraph* ring and are mainly interested in 'the religious aspect'. Their photographer comes round later and spends hours taking photographs, practically demolishing my office in the process. Robin McKie of *The Observer* calls, doing a story that COBE means the universe will end in a Big Crunch. I can't think where he has got this from. I spend half an hour trying unsuccessfully to explain to him that it is nonsense.

I read the electronic mail on my computer. There are messages from Peter Coles, a colleague at QMW who has the office next to mine and who is currently on a visit to the States, giving some further details of the discovery. He has seen the preprints from the COBE team. I send a message telling him what has been going on here. There is also a general mail message with quite a bit of detail from Durham. These messages circulating through the electronic mail grape-vine, show that the COBE team have analysed the possible sources of foreground noise extremely carefully, including telemetry problems, radio interference, the moon, and dust, gas, and cosmic rays in the Milky Way. The image released by NASA shows the whole sky projected onto an elliptical area, with the bright foreground Milky Way emission across the horizontal axis, looking distinctly like the meat in a hamburger. The blotchy patches in the bun of the hamburger are meant to be a highly exaggerated representation of the fluctuations, or ripples. Typically the plus and minus deviations from the average sky brightness are only one part in 100,000 and the variations in density of ordinary matter would have been of the same order. From such inauspicious beginnings did we ultimately evolve, with an average density of about 1 gram per cc, 1,000,000,000,000,000,000,000,000,000,000 (10^{30}) times the average in the universe today.

It turns out, though, that as the maps are quite badly affected by instrumental noise, only some of the blotches will be due to the universe. This does not reduce the significance of the COBE discovery but it is a pity that the NASA press office felt that a picture had to go out

Fig. 12.3 Front page of The Independent, April 24th, 1992.

THE DAILY TELEGRAPH   25/4/92

'Discovery of century if not of all time'

Cosmology v theology

By Christine McGourty, Technology Correspondent

Stephen Hawking
Lucasian Professor of
Mathematics
Cambridge University:
"IT IS the discovery of the
century, if not of all time. It is
thought that there was a
period of very rapid expansion in the early universe.
"This is called the inflationary ... cause the uni-
verse ... bled in
size ...
sec ...

THE INDEPENDENT   25/4/92

Scientists set to work on the ripples in space

THE RIPPLES in space picked
up by Nasa's scientists hold within
them a wealth of clues for ...

By Susan Watts
Technology Correspondent

that first explosion. They rely on
theory to take them closer to the
... Gabib experiment ...

THE INDEPENDENT   25/4/92

Big Bang evidence leaves Christian faith unmoved

LEADING Christian professors
said yesterday that the discovery
of further evidence for the Big ...

By Andrew Brown

confirm or disprove the existence
of God," he said. "Can it modify
theological understanding? ...

Fig. 12.4 Other headlines.

with their press release, even if it is, to say the least, uninformative and perhaps downright misleading. Later, the BBC television programme 'Antenna' was to do a piece on the issue of how the COBE 'ripples' story was covered, which gave the impression that the ripples discovery was not very important and that the coverage of it was pure hype. This programme, in a series about media distortion of science stories, was itself quite a severe distortion of the ripples story.

At 6 pm there is a party at the Planetarium to celebrate the 35th anniversary of BBC Televison's 'Sky at Night', to which my wife Mary and I go. Everyone is talking about the COBE ripples, of course. Patrick Moore, its well-known presenter, tells us that his piece for *The Daily Mail* was written in forty minutes.

Saturday April 25th: *The Independent* quote me as saying that I follow St Augustine and Stephen Hawking in believing time began with the universe and *The Daily Telegraph* reports that I do not believe in God. Stephen Hawking is quoted as saying this is the discovery of the century. I assume that he was trapped into this by being fed the quotes from the press agency report.

Monday April 27th: Down to earth as term starts. I tell the students on my third year Cosmology course about the COBE results and their significance. Later Tim Radford rings to ask if there are any further developments. I say that I have been thinking of writing a more reflective piece about the real significance of the discovery and the press coverage, 1000 words, say. He is interested but says that 'reflective' for *The Guardian* means that it has to be in by noon the next day.

Wednesday April 29th: I fly to Durham to take part in an SERC (Science and Engineering Research Council) visitation to the astronomy group there. A panel of three of us have to assess its work and make recommendations about the level of financial support the group should have. This kind of assessment goes on all the time (my own research grants were similarly assessed earlier in the year). Two members of the group, Richard Ellis, who leads the group, and Carlos Frenk are good friends of mine and part of the QDOT collaboration. Over lunch Carlos and I talk about the significance of the COBE results. He thinks there is a serious problem for the cold dark matter theory, but that there may be a way out. He says that George Efstathiou, at Oxford, another leading member of the QDOT team, favours a variation of the theory in which the universe still has a significant cosmological repulsion term (this is the force which drove inflation in the very early universe, and which has to be very weak today). I say that I am still interested in the possibility of a hybrid universe containing hot and cold dark matter. Carlos is curious but sceptical.

On the Thursday there is a preview of the video of Stephen Hawking's *A Brief History of Time*, organised by *The Guardian* newspaper. At the reception beforehand I ask Stephen what he thinks about the COBE discovery, had he really said it was the discovery of the century? To my surprise he says that he really does think that and that he sees the discovery as confirmation of the inflation scenario. He claims that he predicted the shape of the density fluctuation spectrum, which COBE has observed, from inflation theory. I say what about Harrison and Zeldovich (who had suggested this spectrum about ten years earlier). He says they predicted the spectrum from considerations of how galaxies and clusters form, but *he is not interested in galaxies*. I am a bit stunned by this illustration of how diametrically opposed we are in our world-views.

I find the video quite moving, especially on a large screen. The opening, with the sound of the clicking of Stephen's hand-control as he prepares a text, is very dramatic. The film brings out his concise and oracular style very well. It is also good to see the other relativists who are part of this story, though unhelpful that they are not identified.

The discovery of the century?

Are the COBE ripples the discovery of the century? By no means, but they are a landmark. To set them in perspective, it is worth recalling the other great landmarks in twentieth century cosmology: (1) the discovery of the expansion of the universe by Edwin Hubble in 1929; (2) the invention of the Hot Big Bang model by George Gamow in 1948; (3) the discovery of the cosmic microwave background radiation by Arno Penzias and Bob Wilson in 1965; (4) the explanation of the abundances of the primordial light elements, helium, deuterium and lithium by Bob Wagoner, Willy Fowler and Fred Hoyle in 1972; and (5) the discovery of the motion of our Galaxy through the microwave background radiation in 1977 by George Smoot, David Wilkinson, Francesco Melchiorri and their colleagues at Berkeley, Princeton and Florence, respectively. Of these the Copernican advance is clearly Hubble's discovery of the expansion of the universe. Next in importance comes the discovery of the microwave background radiation. The ripples rank with the others in the above list and, perhaps, others not on the list, in being important details of the Hot Big Bang expanding universe picture.

What are the implications of the COBE team's discovery? First, it needs to be confirmed in other experiments. This seems to have happened already. The COBE results were based on the NASA team's first year's worth of data and the second year's is already 'in the can', being analysed. If these give exactly the same map of the sky, the results will look credible. The Soviet RELICT satellite, which has also been mapping the microwave background radiation, has apparently seen the same result as COBE on the largest scales. A balloon-borne experiment led by Lyman Page of Princeton, Stephen Meyer of Massachusetts Institute of Technology, and Ed Cheng of Goddard Space Flight Institute, has directly confirmed the ripples seen by COBE. Phillip Lubin of the University of California at Santa Barbara has been running an experiment for over a year at the South Pole with a sensitivity approaching that of COBE and seems to have confirmed the discovery, as also has another ground-based experiment by Rod Davies, of Jodrell Bank, Anthony Lasenby, of Cambridge, and their colleagues, at Tenerife in the Canaries.

It will be especially important to look with much bigger telescopes than COBE's, in order to find the much smaller ripples which correspond to nascent galaxies and clusters of galaxies. Because of the limited resolution of the COBE detectors, the fluctuations found so far correspond to structures much larger than anything we can actually map in the universe today. But on these large scales, the COBE team were able to analyse how the strength of the fluctuations declines as we look to larger and larger scales, and the rate of this decline agrees well with what is expected from models based on inflation, like the cold dark matter scenario. Although the agreement between the COBE observations and the cold dark matter model is poor (there is a discrepancy of about a factor of two) the COBE discovery looks good for the inflationary scenario and for a universe dominated by dark matter. And it is certainly great news for Big Bang cosmology. The failure to find these density fluctuations in decades of searching had been one of the most frustrating episodes in modern cosmology.

At a workshop in Erice, Sicily, in August 1992, I asked George Smoot about his feelings concerning the international furore over the ripples. I have known George since 1978, when my family and I spent a year at Berkeley. I recall that in June 1990, during a conference at the Hague in the Netherlands, he and I had dinner at a restaurant on the beach. During dinner, we talked about many things, the anthropic principle, whether there is other life in the universe (I am unusual among astronomers in believing that we are alone), and, of course, whether COBE would detect the ripples. At that time, he was hopeful that over the four years of the mission, they might have a chance of detecting them. I don't think he had any idea of the fame they would bring him. At Erice, George's comment was that the furore had, on balance, been good for astronomy and for physics generally. Wherever he went, he said, astronomers and physicists were telling him that the story of the ripples had helped them in their struggle for resources. I had to agree with him. Considering its significance for our lives, science receives very little coverage on television or in the press, so stories like this are good for science. However, I wish that scientists did not feel it necessary to try to make their work more interesting to the public by talking about 'the mind of God' (Stephen Hawking), or 'the face of God' (George Smoot).

The meaning of the ripples

A few days after the announcement of the discovery, I suggested to my research student Andy Taylor the following question: what can we learn by combining the results of the QDOT survey and COBE? Surely we can derive the whole density fluctuation spectrum on all scales, ranging from the scale of galaxies, where we see large amplitude fluctuations in the mean density (the groups, clusters and voids), to the huge scales studied by COBE, where the density fluctuations are less than 0.01% in amplitude? Andy is well versed in a mathematical technique known as Fourier analysis, which is the key to decoding the primordial density fluctuations. I knew that we had only a few weeks to make any impact on this question. After that there would be a flood of papers from theoretical groups all round the world. On May 26th, only a month after the COBE results were announced, we submitted our paper, which concluded that a hybrid model, containing both hot and cold dark matter, is the best bet. We found that the large-scale IRAS results and the COBE fluctuations require that about 30% of the matter in the universe is in the form of a neutrino species with mass around 7 electron-Volts (eV). 3% of the matter would be ordinary baryonic matter and the remaining 67% would then be cold dark matter, still needed to account for the formation of individual galaxies.

Fig. 12.5 The spectrum of density fluctuations on different scales implied by the COBE ripples and the QDOT galaxy density maps. The solid curve is the prediction of model of the universe containing 3% ordinary baryonic matter, 29% Hot Dark Matter and 67% Cold Dark Matter.

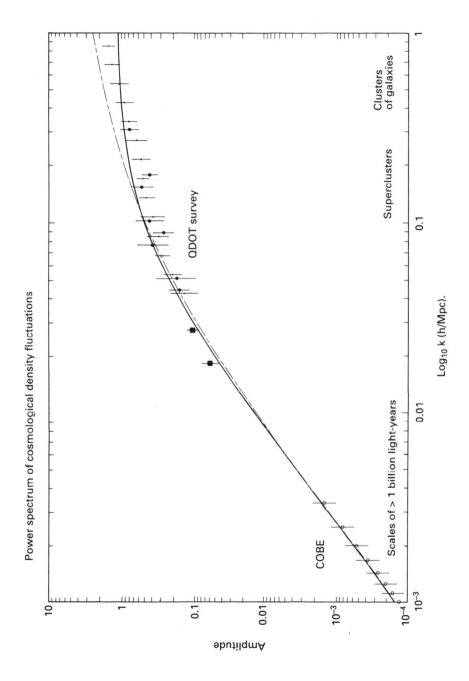

Power spectrum of cosmological density fluctuations

Now there are three types of neutrino known, the electron-neutrino, the muon-neutrino and the tau-neutrino, one for each of the three types of 'lepton' or light particle, the electron, the muon and the tau. Although it is normally assumed that neutrinos have zero mass, some versions of grand unified theories would leave neutrinos with a small non-zero mass. The neutrino masses would be expect to be approximately in proportion to the masses of the corresponding leptons, so the tau-neutrino would be the most massive. Experiments to try to measure these masses have not succeeded yet, and place rather strong limits on the mass of the electron-neutrino. If there is a neutrino species contributing a significant fraction of the mass of the universe it is likely to be the most massive. We therefore concluded that it is the tau-neutrino that has a mass of 7 eV and contributes 30% of the mass in the universe.

A few days after our paper was submitted to the scientific journal *Nature*, we received a paper from Marc Davis and colleagues at Berkeley also submitted to *Nature*, with new simulations of exactly this hybrid model, which show that it can work well in explaining galaxy clustering. Our two papers were published three months later, back-to-back, in the same issue of *Nature*. Later, papers arrive from other American, European and Russian groups arguing for this same mixed hot-and-cold dark matter model. Thus according to many groups of physicists and astronomers, part of the meaning of the ripples is that there may be three kinds of matter in the universe: normal baryonic matter (protons and neutrons) out of which we, the earth, the sun, and the visible parts of galaxies are made; hot dark matter (a neutrino species, probably the 'tau-neutrino', with a mass of about 7 eV); and cold dark matter of a form yet to be discovered. It will be difficult to test directly whether the tau-neutrino has such a mass, for present direct experimental limits on its mass are several mega-electron-volts, approximately one million times higher than Andy and I are predicting. However, there are prospects of indirect tests within a few years through experiments on muon-neutrino beams from nuclear reactors. If neutrinos have a non-zero mass, it is likely that muon-neutrinos will be seen changing into tau-neutrinos, through a process known as 'neutrino oscillations'.

According to some current Grand Unified Theories, if the tau-neutrino has a mass of 7 electron-volts (eV), then the muon-neutrino

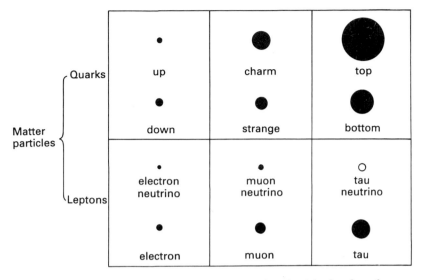

Fig. 12.6 The three typres of neutrino are associated with the three known types of lepton, or light particle: the electron, the muon and the tau. In current Grand Unified Theories of physics, each of these leptons is grouped in a family with two quarks. The six quarks are the building blocks of the baryons (neutrons, protons, etc). The relative sizes of the particles in each row indicate the relative masses of the particles.

ought to have a mass of a few thousandths of an eV. This is just the right mass to solve what is known as the 'solar neutrino problem'. Nuclear reactions in the centre of the sun generate neutrinos and these have been detected in experiments located in deep underground laboratories. The laboratories, which consist of huge tanks of liquid, either very pure water (to detect 'Cerenkov' radiation from the neutrinos) or perchloroethylene (to absorb neutrinos), need to be underground to shield the apparatus from 'cosmic rays', energetic particles which bombard the surface of the earth all the time. Neutrinos pass right through the earth but can be detected when they (very occasionally) interact with the atoms of the liquid in the detection tanks. The problem is that the rate at which neutrinos are detected is only about half the number predicted by models of the interior of the sun. A possible explanation is that one of the neutrino species has a non-zero mass of a few thousandths of an electron-volt. So there is the very interesting prospect that the detection of a 7 eV tau-neutrino would solve the solar neutrino problem, the cosmological dark matter problem, and the large-scale structure problem at one go.

Although several groups have converged on the mixed hot-and-cold dark matter model, this is not the only possibility being advocated. George Efstathiou, with Dick Bond of Toronto, and Simon White of Cambridge, have argued that a universe with cold dark matter and a significant cosmological repulsion can fit both large-scale structure data and the COBE ripples. The problem is that it does not fit in with the mean density of the universe derived from our QDOT survey (see Chapter 9), since in such a model the mean density of the universe would be low. Other theorists are playing with the idea that the form of inflation could be adjusted to change the initial distribution of density fluctuations in the cold dark matter model. Another possibility is that instead of having hot and cold dark matter, we could have hot dark matter together with 'cosmic strings', which are stringlike defects in space-time of enormous energy-density left over from phase transitions in the very early universe. These would then act as the seeds for galaxy formation. Efforts to find strings, for example through the effects of gravitational lensing, have failed so far, however.

Deeper meaning

The deeper meaning of the ripples, though, is that they show that we seem to understand the universe well enough to trace structure back to only 300,000 years after the Big Bang. Radio and microwave astronomers around the world have been trying for decades to find these ripples, to show that we really do know how galaxies and clusters arose from an initially smooth, hot universe dominated by radiation. Even if the COBE ripples do not exactly correspond to the density fluctuations from which galaxies will arise, because they are on much too large a scale, the detection of the smaller-scale fluctuations should not be long delayed now. Seventy years ago we knew nothing of the expanding universe. Fifty years ago we had no concept of the Hot Big Bang and it is less than thirty years ago that we saw the first evidence for such a picture. Now we can say with confidence that we broadly understand the evolution of the universe from a time 300,000 years after the Big Bang until today, thirteen billion years or so later (with lots of details to be filled in). Our understanding of the origin of the light elements shows that we can safely extrapolate this picture back to a time one second after the Big Bang.

With further leaps of faith in current ideas about the early universe, we can trace the origin of structure back far earlier in time, perhaps even to almost the earliest instants in the history of the universe, the moment when inflation began. However, at our present state of knowledge of the early universe, I think it is overstating our understanding rather drastically to say, as some particle physicists are already saying, that the COBE ripples prove that inflation occurred.

The scientific vision 13

WHAT IS SCIENCE? This is a question which many philosophers of science have tried to answer, but their conclusions give only part of the picture. I think this book demonstrates that science today is a complex process, involving imaginative ideas, hunches, guesses, careful observation and experiment, large-scale projects, collaborations and organizations, expensive hardware, painstaking analysis, the long view, instant response at the key moment, determination, luck. Because of this complex reality, scientists tend to have a more practical view of what science is, although many of them are influenced by the ideas of the philosophers. Society in general tends to have an ambivalent view of science, depending on scientists for solutions to technological, environmental and health problems, but regarding science as a dangerous sort of activity, and even as an enemy of human values. In this chapter I will give an outline of some of the views of the philosophers of science, attempt to give a more holistic view of what science is about, and try to respond to some of the attacks that have been mounted on science lately. Science offers a vision of reality which embraces all of life and is central to our culture. But it does not answer all questions and, in particular, it does not provide the meaning of life.

The views of the philosophers of science

Let us start by looking at some of the theories of the philosophers of science. In his influential book *Logic of Scientific Discovery*, Karl Popper advanced the view that science consists of theories proposed by imaginative conjecture, together with attempts to refute these by means of carefully designed experiments. This picture of bold conjectures and austere refutations is one that is easy to defend logically, but it does not take us very far towards understanding how scientists operate. Popper advocates falsifiability as the criterion for calling a branch of study a science. For example, because committed Marxists and Freudians will not agree on any prediction whose failure would lead them to abandon their views, they can not, in Popper's view, be considered scientific.

In *Structure of Scientific Revolutions*, Thomas Kuhn attempts to describe how scientists actually practice science. In what he calls 'normal' periods of a science, there is a scientific paradigm, a communally held set of principles and theories which form the core of the science. These principles and theories are used to develop pictures of widely different situations. Sometimes some peripheral assumptions may be added or discarded to keep the paradigm working. The main activity of Kuhn's 'normal' science can be characterized as puzzle-solving within an established and unquestioned framework. In a 'revolutionary' period, on the other hand, the whole paradigm is overthrown and a new one substituted. Such revolutions include those inaugurated by Copernicus, Newton, Darwin, Planck and Einstein.

These two pictures of the nature of science seem contradictory and there has been much debate on how these two basic pictures, both of which have a clear ring of truth about them, can be reconciled. Imre Lakatos, in his analysis of 'the methodology of scientific research programmes', has tried to incorporate Kuhn's sociological insights into the Popperian viewpoint. The American philosopher Paul Feyerabend amusingly points out that Kuhn's description of normal science could equally well be applied to organized crime. For Feyerabend normal science is science at its least interesting and tends to cultivate dangerous attitudes. He advocates a Trotskyite state of continuous scientific revolution, in which both uninhibited proliferation of new theories and obstinate tenacity to old theories are needed. Everybody may follow their own inclinations and science will benefit from these enterprises.

Feyerabend's philosophical anarchism has strong attractions, especially for the sceptic like myself, but it is a totally hopeless approach for any scientific research programme in which resources have to be acquired by peer review of other scientists. Feyerabend seems to make no distinction between a knowledgeable scientist who holds views considered unorthodox by his or her colleagues (the work of such a person may well benefit science, even if they are wrong), and the crank who knows almost nothing about science but still comes up with what he or she thinks is a theory of the universe. A scientific programme has to be put together in a professional way and has to demonstrate a clear understanding of existing theory. It needs to have a clear goal which must be directed at the unknown. The goal may be more or less fundamental in terms of basic physical theory. Sometimes a programme which starts out with modest goals may turn out to have surprisingly fundamental implications. This is to some extent what happened with our IRAS galaxy surveys.

While Popper may be said to have accurately described the logical position of science (and in essence this is the position taken by Wittgenstein in the *Tractatus*) and Kuhn to have characterized the sociology of science, neither comes close to describing what science means to scientists. What is it, in the minds and imaginations of scientists, that drives them on?

The scientific vision

In my view the essential nature of science – if you like, a holistic view of science – is that it is a vision of the world. And it is only possible to understand what having a vision of the world means if you have experienced such a vision yourself. This could be by entering the imaginative world of a writer or artist. Naturally it is not easy to convey a vision to others who do not share it. The initiation into the scientific vision (a scientific education) is a long and hard road. To outsiders, the community of visionaries may look like a closed religious order and this can evoke hostility. This may be part of the reason for the recent attacks on science as the enemy of human values, and specifically of the Christian religion. To scientists these attacks appear amateurish and wide of the mark. The idea that the scientific vision deprives life of value seems bizarre to scientists, for whom this vision is among the most exciting aspects of their lives.

In spring 1992 several books which attacked science in these terms were published and widely publicised. The launch of these books happened to occur in a week when NASA announced that scientists working with the COBE satellite had detected 'ripples' in the cosmic microwave background radiation and in which the video of Stephen Hawking's *Brief History of Time* was released, so that the pretensions of science in claiming to understand the origin of the universe were seen at their most blatant. Also in the same week a predictably inconclusive debate took place between the Oxford zoologist Richard Dawkins and the Archbishop of York on the existence or otherwise of God. Journalists like Bernard Levin and Auberon Waugh, and the politician George Walden, none of whom appear to be troubled by much knowledge of science (despite the latter having been a minister for science), were quick to jump on the anti-science band-wagon. Are we seeing the same polarization into the two cultures, science and the humanities, highlighted by C.P. Snow over thirty years ago?

The books do not, of course, all take the same line. In *The Creative Moment*, Joseph Schwartz is pro-science at heart, but wants to provide a social critique of science and of the scientific establishment, somewhat in the manner of J.D. Bernal and J.B.S. Haldane. Bryan Appleyard's *Understanding the Present* takes up a much more naive and luddite position. He feels that human history has been ruined because the Christian cosmology of Dante and Aquinas was demolished by Copernicus, Galileo, Newton and Einstein. Mary Midgely, a moral philosopher, provides in *Science as Salvation* the most serious, and therefore in some ways the most irritating, assault of all of these. She has a hidden agenda, which is to make the world safe for Christian theology. All those who deny a role for God, whether physicists like John Barrow and Frank Tippler, who speculate (in their book *The Anthropic Cosmological Principle*) that the purpose of the universe is to populate itself with intelligent life, or biologists like Bernal, Haldane, Monod and Dawkins, are systematically attacked by Mary Midgley for their pretensions. But whereas the physicists present a soft target, speculating about the future in a relaxed moment, I do not find that Mary Midgley comes close to denting the world-view of Jacques Monod's *Chance and Necessity*, that life has arisen in the universe by chance and therefore does not have a grand purpose. You would not guess, from Mary Midgley's pages of attack on Monod, that he is basically giving an account of modern biology and its philosophical implications.

For modern scientists it is possible to see and understand fully only a small part of the vision. Those who study the physics of the very early universe usually know little of the rest of the astronomical universe. Astronomers generally know almost nothing of biology – and so on across the whole of science. It has been said that the last person who knew all of the science of his day was Goethe. Popularizers like James Jeans, Arthur Eddington, Jacob Bronowski or Carl Sagan may try to give us a glimpse of the whole vision, but we know it must be a partial and incomplete one. Yet such popularizations belong in a great tradition that goes back at least to Lucretius's poem *de Rerum Natura* (usually translated as *The Nature of the Universe*). His poem gives a vibrant account of the atomist's vision of the world and is splendidly defiant about the need for gods and religion. The rival Platonic vision, developed into a complete scientific picture by Plato's pupil Aristotle, was far more reverential towards the gods and was easily incorporated in due course into Muslem and Christian theology. We tend to forget that when the works of Aristotle were rediscovered in Europe in the eleventh and twelfth centuries, thereby triggering the European Renaissance, they came accompanied by brilliant commentaries from the Arab philosophers.

Amid the welter of Christian propaganda from the thinkers of the Middle Ages and early Renaissance, we struggle to find something like a modern view. Perhaps only in the writings of Petrarch do we find the perspective that the human being experiencing and questioning the natural world is at the core of what gives life value. Petrarch is not only the source of the romantic movement but he also gives for the first time a vision of nature which is recognizable to the modern scientist.

From the publication of Copernicus's *de Revolutionibus*, the tide which swept the Aristotelean universe away seems to advance remorselessly. Galileo destroys Aristotle's physics, Newton builds a new one, Descartes constructs a new rationalist theory of knowledge. It is not true, incidentally, as Joseph Schwartz suggests in his book *The Creative Moment*, that the mathematization of science dates from Galileo and Newton. Ptolemy and Copernicus were just as mathematical in their treatises. It is just that the mathematics used were different. Ptolemy and Copernicus used Euclidean geometry for their proofs, as did Newton wherever possible, in fact. But the invention by Leibnitz and

Newton of differential calculus was to change the style of scientific work for ever. Today, while much of science can be fully appreciated by anyone with a knowledge of high school mathematics, this is not true for modern developments in general relativity and theoretical particle physics. Even a degree in mathematics is not enough to guarantee being able to make much headway with these. There is therefore a genuine mathematical barrier to understanding some areas of modern science and even most scientists must depend on those inside these fields trying to explain what they are about to the rest of us. Fortunately there are many excellent books by eminent scientists which between them explain almost all aspects of the modern scientific vision.

Why do I find myself here, now?

In the writings of Blaise Pascal, the contemporary of Newton and Descartes, we find for the first time another important strand of modern thought, existentialism. In his *Pensées*, Pascal confronts for the first time the feelings of a human being adrift in the universe brought to birth by the Newtonian revolution.

> *The eternal silence of those infinite spaces strikes me with terror.*

> *When I consider the short extent of my life, swallowed up in the eternity before and after, the small space that I fill or even see, engulfed in the infinite immensity of spaces unknown to me and which know me not, I am terrified and astounded to find myself here and not there.*

These are feelings that have to be grappled with, one way or the other. Pascal himself sought escape from them in religious faith. There can be no scientific objection to such a reaction. Scientists who engage in debates with clerics about the existence of God are wasting their time, because science does not have anything to say about such questions (and clerics who claim they have special insight which allows them to criticize scientific theories are equally mistaken). But Pascal's answer seems to me to be an uninteresting and timid one. The second half of the *Pensées*, grouped together under the heading 'Man With God', is tedious in comparison with the first half, 'Man Without God' (where all the famous Pascal quotations are to be found). Far more impressive to me are those writers, from Hume to Sartre and Monod, who face up to the aloneness of sentient beings in this vast universe as a fact of existence. I do not believe we need the crutch of religion to look the universe in the eye. Nor do the 'new age' fantasies offer anything but an escape from reality. To live one's life in a state of escapism or fantasy is to live only a partial life.

Several scientists have suggested that the meaning of life can be found within scientific theories. Stephen Hawking suggests that this meaning can be found by unravelling the structure of the Big Bang, Paul Davies that it lies in unified theories of physical forces like superstring theory. And a series of scientist-writers, from Freeman Dyson to John Barrow and Frank Tipler, have claimed that a meaning for life can be found in the anthropic principle. My own view, and this is one point where I do agree with Mary Midgley's *Science as Salvation*, is that all these claims are pretentious and superficial.

In the *Tractatus*, Wittgenstein makes a sharp demarcation between what can be said by logic and by science and what can not. All questions of meaning, ethics and value '... we must consign to silence'.

This did not prevent Wittgenstein from spending much of the rest of his life addressing precisely these questions. But it is clear that the answers are as likely to be found in poetry, as they are in the writings of moral philosophers. Personally, though I read the philosophers, what meaning there is to be found, I find in poetry.

What science offers is a vision of reality which embraces all of life, but does not answer all questions. It is superior to alchemy and astrology in its magical powers, that is, in its power to transform nature and predict the future. Of course alchemy and astrology were once scientific visions and the modern vision embraces those aspects of these primitive visions that have permanent value. Human perplexity at the erratic motions of the planets along the zodiac led ultimately to the Copernican and Galilean revolutions. Newton needed the alchemical and hermetic vision to make the step to the concept of a force acting at a distance.

What, then, is the meaning of life? Here I follow Sartre and Monod in asserting that in a very real sense life does not have a meaning. Life and human intelligence evolved by chance. The universe does not have a purpose, though it is self-evident that the universe was such that life could arise in it. Such a statement is far too trivial to merit the title 'anthropic principle'. The properties of other hypothetical universes are of no interest to us.

The meaning seems to drain out of existence if we focus on too large a scale, like Pascal, or on too small a scale like the central figure of Sartre's *La Nausée*, who stares too long at the roots of a chestnut tree. Steven Weinberg, in *The First Three Minutes*, says that life would become meaningless for him if he knew that the universe would come to an end at a finite time in the future. But whether the universe lasts an infinite time or a billion billion years, life seems equally meaningless on those time-scales. Just as we patch together bits of scientific theory into a scientific vision, so we have to patch together meaning out of local fragments, here and now.

On the other hand, life does have value, provided we choose to give it value. As Sartre says, we must be *engagé*. Our oneness with the rest of life on earth, with the stars and galaxies whose evolution made our origin possible, is one of the great themes of modern science. And this theme, the unity of nature, was also at the heart of the poetry of Goethe.

The two cultures revisited

It is above all in human culture, a culture that includes the scientific vision, that we can find at least a fragmentary meaning for life. In 1959 the popular novelist and ex-scientist C.P. Snow caused a stir by suggesting that there were two cultures, that of the literary intellectuals on the one hand and that of the scientists on the other. As a young mathematics student, I remember feeling a certain sense of pride to be identified in this way with a culture comparable to the more overt one exemplified by literature and the arts. The assault on Snow launched by F.R. Leavis in his 1962 *Spectator* article 'The two cultures? The significance of C.P. Snow' came as a shock to me, particularly as at the time I used to make a point of attending Leavis's weekly lecture, rushing from a mathematics lecture to do so. Leavis's article might more accurately have been subtitled 'The insignificance of C.P. Snow', since one of his main points was that Snow had no authority to lecture on the concept of culture.

Re-reading Snow's 'Two cultures' today, I tend to see what Leavis was getting at. Snow has an irritating style, which is simultaneously pompous and garrulous. There is an embarrassing dependence on anecdotal surveys of high table opinion. Snow's main argument seems to be that scientists should be taken more seriously because they are decent types who know a thing or two and who are trying to do some good in the world. In fact there never were two cultures, but only one, of which science is a part. But Snow could be said to have raised an important question: can anyone who has made no effort to know anything about science be said to be cultured? T.S. Eliot and his disciple Leavis would have answered yes, and in doing so led their literary generation into a blind alley. But few serious writers, artists and intellectuals would, I hope, make that answer today.

Resources, rewards and responsibilities

The scientific programmes I have been describing in this book almost all required enormous resources: telescopes, space missions, large accelerators to probe the fundamental nature of matter and of physical forces. These in turn imply large organizations: space agencies like the European Space Agency (ESA) and NASA, observatories, huge accelerator laboratories like CERN. To gain access to these resources, for example to persuade a space agency to build a new astronomical satellite, or to gain time on a large astronomical telescope or particle accelerator, scientists usually have to combine together in large collaborations. Within such a collaboration there might be a range of expertises, from mathematicians and theoretical physicists, to modellers who compare theory with observations or experiments, observers/experimentalists who actually carry out the observations or experiments, and instrumentalists and engineers, who design and construct the equipment needed in the project. It is no easy matter to hold such a collaboration together and to ensure that the right scientific rewards accrue to the diverse members of a collaboration. Much of the detailed work in such a project is carried out by research students and research fellows on short-term contracts, but the credit for the outcome is often attributed to the senior members of the collaboration, who succeed in getting funding for the research students and research fellows, and who direct the scientific programme. Obviously, the credit should be shared between those who have the vision and those who implement it.

Is science an objective search for the truth? This is perhaps an aspect of science that has been overestimated. Scientists try to be honest when presenting their results and there is a great scandal if anyone is found lying about what they have done. But a scientific programme tends to have a certain goal and the scientist has certain expectations about what he or she will find. Scientists do not necessarily feel obliged to point out the weak links in what they have done. The refereeing system for vetting papers prior to their publication is intended to impose the highest standards of explanation, lack of ambiguity, honest attribution to others of their contribution, and to be a filter for genuinely original work. But scientists can only spend a certain amount of their time working on refereeing papers, so these high goals may not always be achieved. Occasionally the anonymous referee system can allow unscrupulous scientists to hold up the publication of the work of their rivals. Personally, I think anonymity is justified in very few cases.

Science is not just about doing good work and getting it published. Little is achieved unless the work is read and used by others. This requires persistent effort in disseminating what one has done to other scientists and to the wider public. Disaster can occur when, as with cold fusion for example, the dissemination to the wider public happens before scientific colleagues have had a chance to read, comment on and test the new discoveries.

One of the points claimed by C.P. Snow in favour of scientists was that by and large they work for the public good. I am not so sure that this is a very strong argument. In Britain, for example, it is estimated that half of all physics graduates go to work in the defence industry. The situation is not so different in other countries. While some of these might be doing so in order to promote the public good, most are simply and not unnaturally going where the well-paid jobs are. Organizations of scientists opposed to the misuse of science for military purposes, like Scientists Against Nuclear Arms (SANA) in Britain (which we recently transformed into Scientists for Global Responsibility), or the Union of Concerned Scientists in the US, attract the support of only a minority of scientists. To be frank it shocks me that so few scientists have a moral concern about how science is used. In a science without much direct practical use for humanity like astronomy, love of the subject and a desire to expand human horizons are certainly a strong part of the motivation. But the desire to be respected and admired by other scientists for doing good work, and perhaps to make a discovery which attracts the public notice, in short the desire for fame, should not be underestimated.

Yet we can not choose whether we live in a scientific age. We do. Scientific studies and perhaps scientific fixes will be needed to ensure the survival of human and other life on earth. Many scientists want to be part of that effort. Not all those who work in the defence industries are happy to be making weapons of mass destruction. It is hard to believe that the solutions to the problems facing the planet can be found without a greater understanding of the scientific vision by politicians, writers, artists, educators, and of course the general public.

Envoi

What, then, is the meaning of the ripples? At one level the lumpiness seen in the IRAS galaxy surveys and the ripples seen in the microwave background radiation by COBE confirm that the universe is dominated by dark matter and that this dark matter plays an essential role in the formation of structure in the universe, from the scale of superclusters of galaxies down to earth itself. When combined with the results from the IRAS galaxy redshift surveys, the ripples may show that in fact there have to be two forms of dark matter, hot dark matter and cold dark matter. The hot dark matter would probably be the tau-neutrino, with a mass of about 7 electron-volts, and this could tie in with a solution to the problem of the deficit of neutrinos from the sun, as well as with Grand Unified Theories which bind nuclear and electromagnetic forces into a single grand unified force in the very early universe. The nature of the cold dark matter remains a mystery, though it is the subject of intense searches at this moment. The ripples may even show that the inflationary scenario for the earliest instants of the Big Bang is correct. In this book I have stressed, however, that there is no direct evidence either for Grand Unified Theories or for inflation.

More generally the ripples show that we have the right general ideas about the history of the Big Bang and how galaxies have formed, even if the details turn out to be different in the future. This is a remarkable achievement of twentieth century cosmology.

But do the ripples tell us anything about the meaning of life? I do not think so. Life on earth, and human life in particular, does not tell us anything about cosmology, beyond the obvious and trivial fact that life had to be possible in this universe. And in turn cosmology does not provide us with any cosmic solution to the problem of finding a personal meaning to life. It is wonderful to study and understand the universe. I find it so anyway and can not resist trying to convince you of this too. But to attach value or meaning to studying the universe is a choice. As Italo Calvino writes in *Time and the Hunter*:

> *Of course, if he chooses, a person can also take it into his head to find an order in the stars, the galaxies ...*

Notes and further reading

2. The lure of number

Further reading:

Ian Stewart, *The Problems of Mathematics* (Oxford University Press, Oxford, 1987).

John McLeish, *Number* (Bloomsbury, London, 1991).

Philip J.Davis and Reuben Hersh, *The Mathematical Experience* (Penguin, London, 1983).

Eli Maor, *To Infinity and Beyond* (Birkhauser, Boston, 1987).

p.7   *Selections from the notebooks of Leonardo de Vinci* (Oxford University Press, Oxford, The World's Classics, ed. Irma A. Richter 1952), p.7 'Let no man who is not a mathematician read the elements of my work.'

p.10   John Fauvel et al. (eds), *Let Newton Be* (Oxford University Press, Oxford, 1988).

p.18   Ludwig Wittgenstein, *Tractatus Logico-Philosophicus*, ed. A.J. Ayer (Routledge and Kegan Paul, London, 1961).

Ray Monk, *Ludwig Wittgenstein, the Duty of Genius* (Jonathan Cape, London, 1990)

Bertrand Russell, *Autobiography* (Bantam, New York, 1970).

p.19   Roger Penrose, *The Emperor's New Mind* (Oxford University Press, Oxford, 1989).

John Barrow, *Pi in the Sky*, (Oxford University Press, Oxford, 1992).
Karl Popper, *The Logic of Scientific Discovery* (Routledge and Kegan Paul, London, 1959).

p.23   K.P. Schaffner, *Nineteenth-Century Aether Theories* (Pergamon, Oxford, 1972).

3. Thirty years of the new astronomy
Further reading:
Michael Rowan-Robinson, *Cosmic Landscape* (Oxford University Press, Oxford, 1979).
John Hey, *The Evolution of Radio Astronomy* (Elek Science, London, 1973).
David Edge and Michael Mulkay, *Astronomy Transformed* (John Wiley, New York, 1976).
Gerrit Verschuur, *The Invisible Universe Revealed* (Springer-Verlag, New York, 1987).
Wallace and Karen Tucker, *The Cosmic Inquirers* (Harvard University Press, Cambridge, Mass., 1986).
Michael Rowan-Robinson, *Universe* (Longman, Harlow, 1990. Published in the US as *Our Universe, an Armchair Guide*, W.H. Freeman, New York, 1990).
Michael Rowan-Robinson, 'Thirty years of the New Astronomy' (*New Scientist*, September 29th, 1990).
Edward Harrison, *Darkness at Night* (Harvard University Press, Cambridge, Mass., 1987).
John D. Barrow and Frank J. Tipler, *The Anthropic Cosmological Principle* (Oxford University Press, Oxford, 1988).

4. Hubble's Law, the cosmological distance ladder and the Big Bang
Further reading:
Edwin Hubble, *The Realm of the Nebulae* (Dover, New York, 1958).
Robert W. Smith, *The Expanding Universe* (Cambridge University Press, Cambridge, 1982).
Michael Rowan-Robinson, *The Cosmological Distance Ladder* (W. H. Freeman, New York, 1985).
Halton Arp, *Quasars, Redshifts and Controversies* (Interstellar Media, Berkeley, 1987).
George Field, Halton Arp, John Bahcall, *The Redshift Controversy*, (W.A. Benjamin, Reading, Mass., 1973)
Dennis Overbye, *Lonely Hearts of the Cosmos* (Macmillan, London, 1991).
Donald Goldsmith, *Supernova! The Exploding Star of 1987* (St Martins Press, New York, 1989).
Joseph Silk, *The Big Bang* (W.H. Freeman, New York, revised edition, 1988).

Stephen Hawking, *A Brief History of Time* (Bantam Books, New York and London, 1988).
Many of the classic papers in twentieth century cosmology can be found in: *Cosmological Constants, Papers in Modern Cosmology*, ed J. Bernstein and G. Feinberg (Columbia University Press, New York, 1986).

5. The years of building IRAS
Further reading:
David A. Allen, *Infrared, The New Astronomy* (Keith Reid, Sheldon, Devon, 1975).
Wallace and Karen Tucker, *The Cosmic Inquirers* (Harvard University Press, Cambridge, Mass., 1986).
Gael Squibb, *IRAS Infrared Astronomical Satellite* (California Institute of Technology, Pasadena, Course AE107, Case Studies in Engineering, 1986).
John Fowler, *IRAS Final Catalog of Pointed Sayings and Small Nebulosities* (Jet Propulsion Laboratory, Interoffice Memorandum, July 23rd, 1984).

p.76 The full IRAS Science Team at the time of launch was:
  US: Gerry Neugebauer (co-chairman), George Aumann, Chas Beichman, Nancy Boggess, Nick Gautier, Fred Gillett, Mike Hauser, Jim Houck, Frank Low, Tom Soifer, Russ Walker, Eric Young
  Netherlands: Harm Habing, (co-chairman), Boudewijn Baud, Douw Beintema, Thejs de Jong, George Miley, Fred Olnon, Stuart Pottasch, Ernst Raimond, Paul Wesselius
  UK: Peter Clegg, Jim Emerson, Stella Harris, Dick Jennings, Phil Marsden, Michael Rowan-Robinson.

6. First view of the far infrared sky
p.90 *The New York Times*, February 22nd, 1983.
p.91 *Sunday Correspondent*, July 29th, 1990.
p.99 *Sunday Telegraph*, November 3rd, 1991.
p.100 *New Scientist*, November 9th, 1983.

7. Discovering the power of IRAS
p.105 *Astrophysical Journal Letters*, special IRAS issue, vol. 278, March 1st, 1984
  The Times, November 10th, 1983
p.107 Results of the QMW-RGO redshift survey were described in *IRAS Sources at High Galactic Latitudes*. A. Lawrence, M. Rowan-Robinson, D.W. Walker, M.V. Penston, R. Terlevich, K. Leech, *Monthly Notices of the Royal Astronomical Society*, vol.219, p.687, 1986.

8. The microwave background story
Further reading:
Steven Weinberg, *The First Three Minutes* (Basic, New York, 1977).

p.110 *Confrontation of Cosmological Theories with Observational Data*, ed. M.S.
    Longair (Reidel, Dordrecht, 1973).
p.111 Michael Rowan-Robinson, Review of *Cosmological Theories with
    Observational Data*, (*Nature*, vol.257, p.520, 1975).
p.112 A. Penzias and R. Wilson, 'A measure of excess antenna at 4080 Mc/s',
    Astrophysical Journal, vol.142, p.419, 1965.
p.115 R.H. Dicke, P.J.E. Peebles, P. Roll and D. Wilkinson, 'Cosmic
    Blackbody Radiation', *Astrophysical Journal*, vol.142, p.414, 1965.
p.117 I am grateful to Anthony Lasenby for drawing my attention to the work
    of le Roux and Shmaonov.
p.118 Michael Rowan-Robinson, 'A sensitive ear for the Big Bang', *Nature*,
    vol.275, p.687, 1978.
    G. Gamow, 'The origin of the elements', *Nature*, vol.162, p.480, 1948.
    R.A. Alpher and R.C. Herman, 'Remarks on the evolution of the
    expanding universe', *Physical Review*, vol.75, p.1089, 1949.
p.124 Michael Rowan-Robinson, 'Aether drift detected at last?' (*Nature*,
    vol.270, p.9, 1978).

9. The IRAS dipole and the 'death' of the Great Attractor
Further reading:
Michael Riordan and David Schramm, *Shadows of Creation* (W.H. Freeman,
    New York, 1991).
John Gribbin, *The Omega Point*, (Heinemann, London, 1987).

p.126 A. Yahil, D.W. Walker and M. Rowan-Robinson, 'The dipole
    anisotropies of the IRAS galaxies and the microwave background
    radiation', *Astrophysical Journal*, vol.301, p.1, 1986.
p.140 M. Rowan-Robinson et al., 'A sparsely sampled redshift survey of IRAS
    galaxies: the IRAS dipole and the origin of our motion with respect to
    the microwave background', *Monthly Notices of the Royal Astronomical
    Society*, vol.247, p.1, 1990.
p.141 T.S. Eliot, The Love-song of J. Alfred Prufrock, *Collected Poems*, Faber
    and Faber, London, 1963.
p.142 *The Independent*, October 7th, 1990.

10. Crisis for cold dark matter?
Further reading:
Bubbles, Voids and Bumps in Time, ed. James Cornell (Cambridge University Press, Cambridge, 1989).
Alan Lightman, *Ancient Light*, (Harvard University Press, Cambridge, Mass., 1991).
Michael Rowan-Robinson, 'Dark Doubts for Cosmology', *New Scientist*, March 9th, 1991.

p.155 W. Saunders, C.S. Frenk, M. Rowan-Robinson, G. Efstathiou,
    A. Lawrence, N. Kaiser, R.S. Ellis, J. Crawford, X.-Y. Xia and
    I. Parry, *Nature*, vol.349, p.32, 1991.
    In December 1992, our *Nature* paper became, briefly, the most cited paper in Physics, according to the *Science Citation Index*.
p.156 *The New York Times*, January 3rd, 1991; *Financial Times*, January 4th, 1991; *The Daily Telegraph*, January 4th, 1991; *Evening Standard*, January 4th, 1991.

11. From quasars to ultraluminous infrared galaxies
Further reading:
Michael Rowan-Robinson, 'The Great Quasar Odyssey', *New Scientist*, November 4th 1982.
Halton Arp: *Quasars, Redshifts and Controversies* (Interstellar Media, Berkeley, 1987).

p.161 C. Hazard, M.B. Mackay, A.J. Shimmins, 'Investigation on the radio source 3C273 by the method lunar occulation', *Nature*, vol.197, p.1037, 1963.
    M. Schmidt, '3C273: a star-like object with large red shift', *Nature*, vol.197, p.1040, 1963.
p.167 M. Rowan-Robinson et al., 'An IRAS galaxy of phenomenal luminosity: protogalaxy or embedded QSO?' *Nature*, vol.351, p.791, 1991.
p.177 *The Times*, June 27th, 1991; *The Daily Telegraph*, June 27th, 1991; *Daily Mirror*, June 27th, 1991.

12. COBE discovers the cosmic ripples
p.184 *The Independent*, April 24th, 1992.
p.186 *The Daily Telegraph*, April 25th, 1992; *The Independent*, April 25th, 1992.
    G. Smoot et al. (*Astrophysical Journal Letters*, vol.396, p.L1, 1992).
p.181 The Press Agency report which triggered the UK press coverage in fact

originated from George Smoot via a San Francisco freelance science writer.

p.187 Michael Rowan-Robinson, 'Yesterday's ripples, today's galaxies', *The Guardian*, May 1st, 1992.

p.190 A. Taylor and M. Rowan-Robinson, 'The spectrum of cosmological density fluctuations and nature of dark matter', *Nature*, vol.359, p.396, 1992.

　　　M. Davis, F.J. Summers and D. Schlegel, 'Large-scale structure in a universe mixed with hot and cold dark matter', *Nature*, vol.359, p.393, 1992.

13. The scientific vision

p.198 Karl Popper, *The Logic of Scientific Discovery* (Routledge and Kegan Paul, London, 1959).

　　　Karl Popper, *Conjectures and Refutations* (Routledge and Kegan Paul, London, 1963).

　　　Thomas Kuhn, *The Structure of Scientific Revolutions* (Chicago University Press, Chicago, 1962).

　　　Imre Lakatos and Alan Musgrave (eds), *Criticism and the Growth of Knowledge* (Cambridge University Press, Cambridge, 1970).

　　　Paul Feyerabend, *Science in a Free Society* (NLB, London, 1978).

p.200 Mary Midgley, *Science as Salvation* (Routledge, London, 1992).

　　　Bryan Appleyard, *Understanding the Present* (Picador, London, 1992).

　　　Joseph Schwartz, *The Creative Moment* (Jonathan Cape, London, 1992).

　　　John D. Barrow and Frank J. Tipler, *The Anthropic Cosmological Principle* (Oxford University Press, Oxford, 1988).

　　　Jacques Monod, *Chance and Necessity* (Collins, London, 1970).

　　　Lucretius, *On the Nature of the Universe*, translated by Ronald Latham (Penguin Classics, London, 1951).

p.202 Blaise Pascal, *The Pensées*, translated by J.M. Cohen (Penguin Classics, London, 1961).

p.204 Jean-Paul Sartre, *La Nausée* (Gallimard, Paris, 1938; translated as *Nausea*, Penguin, London, 1965).

　　　C.P. Snow, *The Two Cultures and A Second Look* (Cambridge University Press, Cambridge, 1964).

　　　F.R. Leavis, *The Two Cultures; significance of C.P.Snow?* (*Spectator*, 1962).

p.208 Italo Calvino, *Time and the Hunter*, translated by William Weaver (Abacus, London, 1983).

Footnote: on June 23rd 1993, a British mathematician, Andrew Wiles, announced at a lecture in Cambridge that he had proved Fermat's last theorem (see p.15). The complete proof is about 200 pages long.

Acknowledgements for figures

Frontispiece   K. Matthews, T. Soifer, J. Larking, California Institute of Technology/CARA
4.1, 4.5a   Observatories of the Carnegie Institution, Pasadena
4.2   Lick Observatory
4.4b, 4.7b, 11.4b   Mt Palomar Observatory, California Institute of Technology
4.5b   Gerard de Vaucouleurs, University of Texas
4.7a, 9.5a, b   National Optical Astronomy Observatory
4.8b   Alan Stockton, University of Hawaii
4.10   Alan Guth, Massachusetts Institute of Technology
5.1, 5.3, 5.4, 6.1a, 6.2a, 6.3, 6.4, 7.1, 7.4a, 12.1   National Aeronautics and Space Administration
5.2a   Gerry Neugebauer, California Institute of Technology
5.2b   Frank Low, University of Arizona
6.1b   *The New York Times*
6.2b   *Sunday Correspondent, Los Angeles Times*
6.5   *Sunday Telegraph*
7.2a   Tom Chester, California Institute of Technology
7.2b   Tom Soifer, California Institute of Technology
7.3   Mark Jones, Open University
7.4b   *The Times*
8.1, 8.2, 8.3   Arno Penzias, AT & T Bell Laboratories
8.5a   John Mather, Goddard Space Flight Laboratories
8.6, 8.7, 12.2   George Smoot, University of California, Berkeley
9.3   (a) Carlos Frenk, (b) George Efstathiou, (c) Nick Kaiser, (d) Will Saunders,
(e) John Crawford, (f) Andy Lawrence
9.7   *The Independent On Sunday*
10.2c   Jim Peebles, University of Princeton
10.3a   George Efstathiou
10.3b, c   John Huchra, Center for Astrophysics, Harvard
10.6   *The New York Times, Financial Times, Evening Standard, The Daily Telegraph*
11.2, 11.7   National Radio Astronomy Observatory
11.5   Royal Greenwich Observatory
11.9   *The Times, The Daily Telegraph, Daily Mirror*
12.3   *The Independent*
12.4   *The Daily Telegraph, The Independent*

Index

217